SEED OF KNOWLEDGE, STONE OF PLENTY

SEED OF KNOWLEDGE, STONE OF PLENTY

UNDERSTANDING THE LOST TECHNOLOGY OF THE ANCIENT MEGALITH-BUILDERS

JOHN BURKE AND KAJ HALBERG

Council Oak Books
San Francisco/Tulsa

First edition, first printing, 2005
Printed in Canada

Photography by Kaj Halberg, unless otherwise noted
Cover design by Lightbourne
Interior design by Melanie Haage
Technical drawings by Carl Brune

ISBN 1-57178-184-6
ISBN 978-1-57178-184-0

Library of Congress Cataloging-in-Publication Data

Burke, John (John A.)
Seed of knowledge, stone of plenty : understanding the lost technology of the megalith builders / John Burke and Kaj Halberg.
p. cm.
Includes bibliographical references and index.
ISBN 1-57178-184-6 (hardcover)
1. Megalithic monuments. I. Halberg, Kaj. II. Title.
GN790.B85 2005
930.1'4--dc22 2005018986

To Helena and Judy

Table of Contents

Acknowledgments

First and foremost, we wish to thank Dr. "Lefty" Levengood, whose work was vital to this volume.

For information, comments, suggestions, lending us photographs, or helping us in different ways, our warmest thanks go to the following persons: Dave Barron, Nancy Burson, Judy Chiang, Dr. Bruce Cornet, Carol and Bill Cote, Jiri Dvorak, Starr Fuentes, Dr. Geoff Groom, Rosemary Ellen Guiley, Lois Horowitz, Phil Imbrogno, James and Shelley Keel, Søren Lauridsen, Dr. Brad Lepper, Glenna Levengood, Alex Patterson, Roberta Pulhalski, Ed Sherwood, Roslyn Strong, Nancy Talbott, and Anastasia Wietrzychowski. And to Paulette Millichap, Ja-lene Clark, Laura Wood, Rebecca Loftiss, and the whole family at Council Oak Books.

Introduction

THE MISTS HANGING HEAVILY OVER THE RAIN FOREST CANOPIES OF MESO-America are the mists of time, for they serve as a nebulous shroud for peoples whom time has forgotten. Only the peaks of their greatest creations – towering limestone pyramids – pierce the shroud and whisper of life. At many other places around the world, huge earthen mounds or stone structures are silent sentinels from times, cultures, and knowledge long since gone – in Illinois, France, England, Bolivia, and Egypt, to mention but a few. Today, these structures beckon us to them. When were they built? Who built them? And why? Archaeologists have long since revealed when and who. The why, however, has often been guesswork.

One of the hardest parts of investigating something built before writing was developed is trying to find evidence of how the people who built it actually used it. The literature of Stonehenge and hundreds of other ancient megalithic sites usually states that these structures were used for ceremonial purposes, probably of a spiritual nature. However, the 'ceremonial site' label is simply an interpretation that has, over time, become enshrined as fact. Among academics working in the field, no one could think of any practical use for a Stonehenge or a pyramid. So, if they were devoid of practical purpose, they must have been used only for ceremony. This reasoning has become so ingrained in our view of prehistory that these structures are often referred to as 'sacred sites.' We need to remember that we view these sites through the tinted glasses of our own culture, which divorces the spiritual from the practical.

For example, consider the twentieth century's largest structures. Hydroelectric dams are probably the biggest structures humanity has built to date. And why did we build them? Even someone who had no acquaintance with turbine-generated electricity could surmise that these structures are important to our society by the effort we put into them. We know very well why we are willing to invest huge amounts of money, labor, and time in erecting these dams. We made this very physical effect because the return was worth it, in very physical terms. From these dams we gain electricity, the lifeblood of an industrial civilization. What if our pre-industrial ancestors also invested huge amounts of labor and time to erect enormous creations of stone and earth because it quite simply was worth it, in physical terms? What if the pyramids, mounds, and henges paid their builders back by producing fertility, the lifeblood of every agricultural civilization? In many cases, we have well-documented evidence that these structures were dedicated to fertility gods or contained symbols and tokens associated with fertility, but they may have actually worked as mechanisms for increasing crop yields.

What if you knew that many of these monuments do in fact produce physical effects, even today? What if you knew that they were built on ground where certain natural electromagnetic energies are concentrated, and designed in such a way as to further concentrate these energies? Finally, what would you say if you knew that pyramids, henges, and mounds were usually built only after a food crisis arose, and that the way they concentrated these energies had the end result of producing more food?

Consider the following facts:

- Megalith building seems to have begun in each country only after there was a crisis of agricultural productivity, and famine loomed. The builders of mounds, pyramids, and henges were often fighting for survival when construction began, yet archaeological evidence shows they got wealthy soon after the buildings had been completed.

- Experiments by academics in Europe have shown that the slash-and-burn agriculture, known to be employed when these megaliths and mounds were built, will exhaust the soil in three years. Yet ancient European farmers somehow got satisfactory production for seven years or more in the same fields, without the as yet-to-be-discovered help of fertilizer or crop rotation. Experts know that it was done, but cannot explain how. Something similar was true for the Mayans in Meso-America. Their agriculture fed millions of people in the Yucatan Peninsula, which today can barely support a hundred thousand. Before the Inca, a little-known Andean culture seems to have tapped earth energies to produce a similar effect, growing bountiful crops in the harsh altiplano, where farmers struggle to get by today.

- In England, excavations at causewayed enclosures and henges showed that emmer wheat had been carefully cleaned of all weed seeds before being brought to the site and placed at the causeway's ditch. This was wheat as seed, not as food. Throughout Europe, for a thousand years such enclosures were sited on ground above subterranean geological structures that generate natural electrical ground current, of the same kind we found linked to improved seed performance on Mayan pyramids in Guatemala.

- In North America, hundreds of mounds were built by tribes of the loose-knit group called the Mississippian Culture. Most of these mound-building peoples vanished long before Europeans arrived. However, during the white settlement age, the Natchez tribe still used mounds, and in 1730 a French Jesuit missionary wrote home to his superior to say that no Natchez farmer would dream of planting his seed without first bringing it to the top of the mound for certain 'blessings.' Something similar was true for the Aztecs.

- In the original American mound-building culture, the Olmec of Mexico, villages with mounds enjoyed a higher standard of living than otherwise identical villages without mounds a few miles down the same river.

- Today, Mayan farmers still bring their seed to the top of certain pyramids in Guatemala.

- Our own experiments, which we invite you to copy, have shown that seed of ancient varieties, which are still produced today, if left for a time in the air at such ancient structures, often grows faster and more vigorously, while producing up to double or triple the amount of food. Corn seeds, placed by us on one of the oldest Meso-American pyramids, grew dramatically better, particularly if placed there on days of high electric energies. Seeds that we placed on North American Indian mounds showed dramatically improved growth, especially when lightning storms were nearby.

- Twenty-first century seed treatments using contemporary versions of the same electrical energies present at the megaliths have achieved

> the same effects that we have observed in seed placed at those ancient sites: faster growth, higher germination percentage, better stress tolerance, and higher yields. These results have been confirmed many times by universities and agricultural organizations.

You need not take our word for these facts; you can confirm them for yourself today, without so much as a trip overseas. These structures are so widespread in America that two-thirds of the population live within a few hours drive of one. Anyone who wishes can confirm or refute our findings with the information provided in this volume.

My research has taken Kaj and me on a journey to remote places and times. In the following chapters, we shall take you to such sites in both North and South America. We shall look at how these energies arise everywhere from natural forces, how they could have been detected by the ancient builders (or you), and how these energies affect seed in a way that increases food production. Then we shall travel back in time to Europe and beyond to see how generations of archaeologists have unearthed mountains of evidence that are entirely consistent with this new understanding of ancient technology. Take the journey with us, and you be the judge.

—JOHN BURKE, *July 2005*

CHAPTER 1

The Lost World

"However, other Mayan areas, such as the well-studied cities of Copan and Tikal, show little archeological evidence of terracing, irrigation, or raised or drained field systems. Instead, their inhabitants must have used archeologically invisible means to increase food production."

—JARED DIAMOND, *COLLAPSE: HOW SOCIETIES CHOOSE TO FAIL OR SUCCEED*

IN THE DARK TROPICAL NIGHT, THE DENSE GUATEMALAN RAIN FOREST OF TIKAL loomed over us. Decaying vegetation emitted a distinct smell, mixed with the fragrance from flowers and herbs. Insects called incessantly, and a coughing roar announced that a jaguar was on the prowl.

Even at 3:30 in the morning, our clothing was drenched in sweat and stuck to our bodies. Head lamps on, instruments in hand, Kaj and I wound single file through the undergrowth up the jungle trail. We had been walking uphill at top speed for thirty minutes behind Luis, our guide. To catch our breath, we sat down on a wall between the famous King's and Queen's Pyramids (Plate 1), brooding silently in a moonlit fog. Re-entering the pitch-dark rain forest, we climbed the winding trail, emerging onto a small plateau known as *El Mundo Perdido*—The Lost World. At this moment, the readings of airborne electric charge, recorded by our electrostatic voltmeter, suddenly leapt way beyond anything we had ever measured before.

With the deep-throated roars of howler monkeys surrounding us in the pre-dawn darkness, we watched with some alarm the already striking readings growing even stronger as we approached The Lost World Pyramid (Plate 2), then rising again as we ascended its oversize steps. In a flash, we realized that our hunch had been right.

We had come to Guatemala toward the end of the second millennium AD, searching for answers to a question dating back to the first millennium BC. Archaeologists have done a very good job of figuring out who built the Mayan pyramids, as well as when and how. The question that we had come to seek the answer to was why? We thought we knew. Now we were trying to find hard evidence, evidence that would stand up to scientific peer review.

We were equipped with the electromagnetic instruments that had served us so well at many other ancient sites around the world, from English henges and mounds to Native America's mysterious rock chambers and the biggest earthen mounds in the world. The instruments we had applied at all these sites were similar to those used by the U.S. Geological Survey. Time and again at ancient structures, the instruments had revealed unusual concentrations of geo-magnetism, electrical ground currents, and electric charge in the air. A few other pioneers had noticed this before us, though never at so many different locales.

A review of previous research, along with site visits with our instruments, began to show that ancient farming civilizations had repeatedly selected spots where natural electrical energies were strongest. There, they had invested mind-boggling amounts of labor to build structures whose design further concentrated these natural energies.

More years in research libraries surveying archaeological findings revealed a pattern. The megaliths of various forms were not built when you would expect it. If these pyramids, henges, mounds, etc., were purely symbolic monuments celebrating something, you would expect them to be built when a civilization was in its prime and had resources to spare. In fact, following the histories of these sites in chronological order generally showed the opposite to be true. Although it sounds suicidal, these labor-intensive behemoths repeatedly were built at a time when the available land had become exhausted through overuse – in the days before fertilizer and crop rotation. With a food crisis at hand, a society would suddenly take up to twenty-five percent of its work force and put it in a multi-year (or even multi-decade) project, building an enormous structure with no apparent practical value.

You would expect these societies to at least then continue their slide into poverty and hunger. Yet, the opposite occurred again and again – far too often to be mere coincidence. Once these pyramids, mounds, or rock chambers were completed, the society would suddenly start to prosper. There is some missing factor regarding these ancient structures, something crucial that archaeologists are not aware of. There are good reasons to believe that this had to do with tapping the earth's naturally occurring electrical energy to produce more food, in a manner not that different from a modern technology that does the same thing today.

THE PEOPLE OF THE VOLCANO

High civilization first arose in the Americas in the most unlikely of settings. Vera Cruz province on Mexico's Gulf Coast is even today so inhospitable that its twenty-thousand square miles of mosquito-filled swamps contain few villages, fewer roads, and can test even the most inveterate traveler. Yet, over three thousand years ago, something remarkable happened here to a people about whom we know next to nothing, not even what they called themselves. In recent centuries, as Mexicans began to stumble on mysterious ruins hidden deep inside uninhabited jungle, they simply began referring to them by the only other thing of value that was found there: rubber trees. And so this long-vanished population became the *Olmec.*

In a land of flat, featureless terrain half again as large as Connecticut, Rhode Island, and Massachusetts combined, lies a single 'oasis' of the vertical: the lonely, mist-shrouded volcanoes of the Tuxtla Mountains. Here, rocky ridges arc gracefully skyward, pulsing with an invisible electrical force – a force the Olmec may have taken with them to new lands.

At another volcano, Piton de la Fournaise – off Madagascar on the island of Reunion – French scientists have measured how rainwater running off through underground channels in volcanic rock will generate electrical charge and magnetic fields, usually concentrated at the highest points.[1] Other geologists have captured such charge on instruments atop sacred Mexican volcanoes, such as Popocopetol near Mexico City. At 'Popo,' the electric charges reach thousands of volts per meter.[2] Did the ancient Olmec wish to export this effect from their homeland? After farming the Tuxtla slopes for centuries, they moved out into the surrounding swamplands and took with them enormous quantities of their local basalt, volcanic lava that has been hardened under heat and pressure. One of the best examples today is a remote and uninhabited plateau, named after a nearby village, San Lorenzo. After a trying journey by mule through tropical flatlands, full of biting insects, we came upon a 160-foot rise, ascending to a half-mile long, flat top where over two hundred small mounds were built circa 1250 BC, centuries before the founding of Rome.

Archaeologists had long assumed that this small plateau was part of the natural landscape. The famous excavator Michael Coe, however, found something very exciting when his team had cleared the summit of growth and began to dig in earnest.[3] This flat piece of ground, standing sixteen stories above its surroundings was not entirely natural after all. At least the top twenty-five to thirty feet were composed of earth, carried up here – basket by basket – by its Olmec founders. A good portion of it was not just any kind of dirt, but layers of an unusual type of gravel, mined from stream beds. It forms a distinctively different color from the earth above it and below it, because the pebbles are stained with iron. This would make them highly electrically conductive, and most of the two hundred small mounds on the top rested on a pile of the pebbles. These weren't the only large-scale artificial features. Steep-sided, knife-edged ridges were built to run out and down from the plateau, looking every bit like the volcanic basalt ridges back home in the Tuxtla volcanoes.[4]

In fact, there is a great deal of Tuxtla basalt built into the San Lorenzo plateau in such a manner that it may have acted as veins of electrical current pulsing within the hill, created by water coursing through rock, much as on the volcanic slopes of the Olmec homeland. No fewer than fourteen springs at the base of the hill are connected to twenty-some man-made lakes atop the hill, lakes lined with water-repellent *bentonite* blocks, a type of magnetic stone.[5] Sluice gates were built into the lakes that, when lifted, sent water rushing down through the hill inside a series of man-made drains, composed of hundreds of tons of tightly fitted, quarried blocks of basalt that had been transported on rafts through sixty miles of swamps from the Tuxtla Mountains. Herculean labor was expended in an extremely forbidding place. Why?

One characteristic of rushing water that intrigues us is its ability to generate electric charge. This effect can be dramatically illustrated with a simple device, called a Kelvin water dropper. Start by placing an LED indicator light bulb (like the small green 'on' light on a computer) between two separating streamlets of water. The build-up of electrical charge, generated by the water droplets, will actually cause the indicator bulb to light up about every twenty seconds.[6,7] Two special characteristics of the Tuxtla basalt are of particular interest here. Because of its high content of magnetite and other metals, this rock is fairly magnetic. Secondly, the high metal content[8] makes it an efficient conductor of electricity. Furthermore, the ability of any rock to conduct electricity is proportional to its water content. If a sluice gate at the main reservoir on top of the San Lorenzo hill was opened, the water would rush down the drains, through the interior of the hill. Now, the rocks of the drains had all the above properties, making them electrical 'veins' to carry the charge of the running water throughout the hill. The electric charge would concentrate its strongest effects on the mounds on top (just as ground charge accumulates at high points during a thunderstorm), and the knifelike ridges would conduct more current up from the jungle floor below. What were they hoping to accomplish?

RICHER THAN THEIR NEIGHBORS

In a pattern that we shall see repeated again and again over the continents and millennia, this awe-inspiring building effort followed poverty and preceded riches. The Olmec were originally subsistence farmers, practicing slash-and-burn agriculture. When their numbers swelled beyond what their homeland could support, they fanned out into the surrounding lowlands. But solid ground was scarce in this region where most acreage is under water year round and most of the rest is flooded during the rainy season. So they farmed what was left: the tops of the natural levees beside the riverbanks.[9] These were narrow but fertile strips, their soil renewed by mud deposited in the floods, much as in the Nile River Valley a world away in Egypt. Still, full-time farming has always led to the same problem: runaway population growth. The narrow levees soon grew crowded, and pressure continually grew on the ability to feed more people on a fixed amount of land.

The levee settlements developed into two basic patterns: sites with artificial earthen mounds and sites without such mounds. In *Science,* William F. Rust and Robert J. Sharer describe a puzzle regarding these settlements: Villages with a mound always fared much better than virtually identical ones a few miles down the same river without a mound.[10] Analyses of trash heaps and skeletons showed that mound villagers enjoyed a significantly higher standard of living. They ate more meat, for example, and otherwise lived healthier lives. They had a much higher social status and were far wealthier, possessing many valuable items, such as polished serpentine tablets and ornaments of jade.

As a whole, the Olmec began to prosper in their new homeland and grew sophisticated indeed. Imported magnetite was polished into concave mirrors that could focus the sun's rays like a reflective version of a magnifying glass, starting fires. As they polished these stones, the Olmec must have noticed how doggedly the dust would cling – a result of magnetic attraction. In fact, a needle of this magnetic ore (magnetite) was excavated by Michael Coe at San Lorenzo, complete with a groove for suspending it from a string, an arrangement that would allow it to act as a compass needle. Because the magnetite ore was formed when the magnetic poles of the Earth pointed a slightly different direction, their needles point eight degrees west of present-day true magnetic north. Intriguingly, many Olmec and Mayan structures are strictly aligned with an axis that points eight degrees west of today's magnetic north.[11] If in fact this is what it looks like, then the Olmec invented the compass a thousand years before the Chinese.[12]

No one can figure out how the Olmec paid for their many luxury objects. Nothing has been found that they were exporting in turn. Some speculate that perhaps they exported perishable handicrafts, or the brightly colored feathers of birds, none of which would turn up in a modern dig. We began to wonder if the wealth may have been based on something else perishable – food. Surplus agricultural production can always be easily exported in return for luxury goods, and it fails to show up in excavations.

At San Lorenzo, over time, bigger and bigger blocks of basalt were brought in. Weighing up to thirty tons, they were floated up to sixty miles through swamps and dragged to the top of this partly artificial hill. Some of them were carved into human heads, eight to ten feet across. Others were carved into blocks, flattened on top and decorated around the sides with fertility symbols, including the ubiquitous were-jaguar, half man and half cat. Wear patterns on the tops show that something was frequently placed on them, something heavy enough to have worn the hard basalt at the edge and center of the flat-topped surface.

In Western Guatemala, similar basalt sculptures demonstrate a clear ability of the Olmec to determine the north and south poles of magnetic fields. If a whole body was depicted, the magnetic poles straddled the navel. In the seven large basalt heads, there were detectable north poles located at the right temples of the heads. These were not inserts in the heads; rather, these poles were present in the original rock. This positioning is obviously not by accident and suggests the carvers may well have used an Olmec lodestone compass to detect magnetic polarities in the basalt and then carved the sculpture accordingly[13,14] In a later chapter, we will see such ancient abilities on display again in English henges.

It is known that after completing the hill of San Lorenzo, the Olmec prospered. At La Venta, 100 miles to the northeast, they later built an even more impressive set of fertility images with basalt structures, including an imitation volcano. Unfortunately, La Venta today is situated atop the richest oil deposit in Mexico, and is not open to the public. The ruins lie on an island in the middle of a volcanic lake that also sits precisely atop one of the largest gravity anomalies in Mexico, a correlation with another geophysical force that we discuss in later chapters. Further west, at Tres Zapotes, these impressive Olmec geo-engineers leveled the top third of a mountain and erected on it a series of four-sided stone pyramids that were flat on top. This creation was at the peak of a long mountain ridge from which much of their magnetite was mined, and would therefore be littered with magnetic anomalies.

END OF AN ERA

The Olmec disappeared into history after a sustained drought. Less rain fell, so the rivers flooded less frequently, depositing less fertilizing alluvium and leading inevitably to soil exhaustion. The people's response was dramatic. The huge, carved basalt heads were ripped from their perches and buried, but not just anywhere. They were dragged out onto the knife-edge ridges at great risk and interred there under thin layers of soil.[15] Was this a desperate attempt to increase the conductivity of the ridges, drawing up natural telluric current, thereby reinforcing the charging effects of the drains?

Heads and altars (all the extra available basalt) were also buried on the plateau top, but again, not just anywhere. They were placed only in an east-west line directly above the chief underground drain. The faces of the heads were disfigured before burial, so it seems that any ceremonial aspect of these creations had been abandoned. Do these remains hint of a heroic, yet ultimately futile, effort to survive?

We do know one thing they did, shortly before vanishing into the mists of time. They conveyed some of their knowledge to a group of rainforest dwellers, whom we today call the Maya. At a spot loaded with geologically induced electric ground charge, the Olmec brought their influence to bear in the creation of their last pyramid and the Mayans' first: the very special structure now known as the Lost World Pyramid.

In Chapter Four, we shall return to look at the startling results of our tests there – results that show profound changes in agricultural growth produced by the energies harnessed at the Lost World Pyramid. But first, let us explore the nature of these natural energies and how ancient builders might have detected them and harnessed them in the first place.

CHAPTER 2

Harnessing Nature's Electromagnetic Energy

"Any sufficiently advanced technology is indistinguishable from magic."
— Isaac Asimov

We live our lives engaged in a daily electromagnetic dance with our planet. Earth is hardly a stable world. In fact, it pulses every day with powerful rhythms of electrical and magnetic force, and so do we. Earth produces a magnetic field much like a bar magnet. Any compass needle tells us that the north pole of our earth magnet is near the physical North Pole, which is one end of the axis on which our planet rotates. Life on Earth would be impossible without this field, called the *geomagnetic field.* The geomagnetic field deflects the solar wind, deadly blasts of electrically charged high-energy particles from the sun. Mars lacks a magnetic field, causing its surface to be hostile to life. But the geomagnetic field takes a beating doing its job. The field is depressed when struck by 'gusts' of solar wind, much like a warrior whose shield deflects a mighty enemy sword strike but recoils in the process. An aurora can be produced by an unusually powerful solar gust, and is roughly analogous to the ringing of the shield under a particularly vicious blow. During the Northern Lights, air molecules at the upper edges of the atmosphere are so excited by the impact of solar wind that they glow.

When our part of Earth rotates into sunlight at dawn, the geomagnetic field recoils from the impact of solar wind, and this affects the field lines. Field lines can be thought of as linear incarnations of the magnetic field. Sprinkle iron filings on a piece of paper above a magnet and you will watch the filings arrange themselves along these invisible lines of force. At dawn, the magnetic field lines shrink (Figure 1), which makes them stronger. That means that the strength of the geomagnetic field running through the land, our homes, our bodies, and our brains surges each dawn. Conversely, at night the geomagnetic field lines are no longer being compressed by solar wind, and they gradually stretch into a long tail emanating from the dark side of the planet in a pattern reminiscent of a comet. This lengthening of the field lines weakens them. The end result of all this is that the geomagnetic field weakens at night only to come roaring back quickly as dawn approaches. There are places where the local geology makes this effect stronger than at others due to the principles of electromagnetism.

'Electromagnetism' is a single word for a reason. Magnetism and electric force are inextricable twins. A moving electric current generates a magnetic field, and a changing magnetic field generates electric current in anything present that will conduct it. This is how our electric power plants work. Physical force from coal, oil, or falling water moves a mass of copper wires past a huge magnet, and an electric current is generated. This is the principle of physics known as *induction.*

Earth itself is subject to these same forces. When dawn brings a change in magnetic field strength, it actually generates weak DC currents in the ground. Like all electric currents, these *telluric currents* travel better in some media than others. Ground with lots of metal or water within it conducts these natural, daily currents particularly well. Drier or less metallic ground conducts it less well. When these two types of land intersect, we have what geologists call a *conductivity discontinuity*, and interesting things happen there. The ground current hitting this boundary has a tendency to either reinforce or weaken those daily magnetic fluctuations – sometimes by several hundred percent.[1] This change in magnetic field strength in turn generates more electric current. So conductivity discontinuities are 'happening places.' Their magnetic fluctuations and ground currents are much higher than in surrounding areas. It is our good fortune that it is the z-axis, the axis that our magnetometer measures,[2] of the geomagnetic field that is affected this way.

One important effect of these ground currents is that they will attract electrified air molecules of opposite sign. A positive electric current in the ground will draw negatively charged air molecules toward it and vice versa. These effects are magnified on islands or peninsulas.

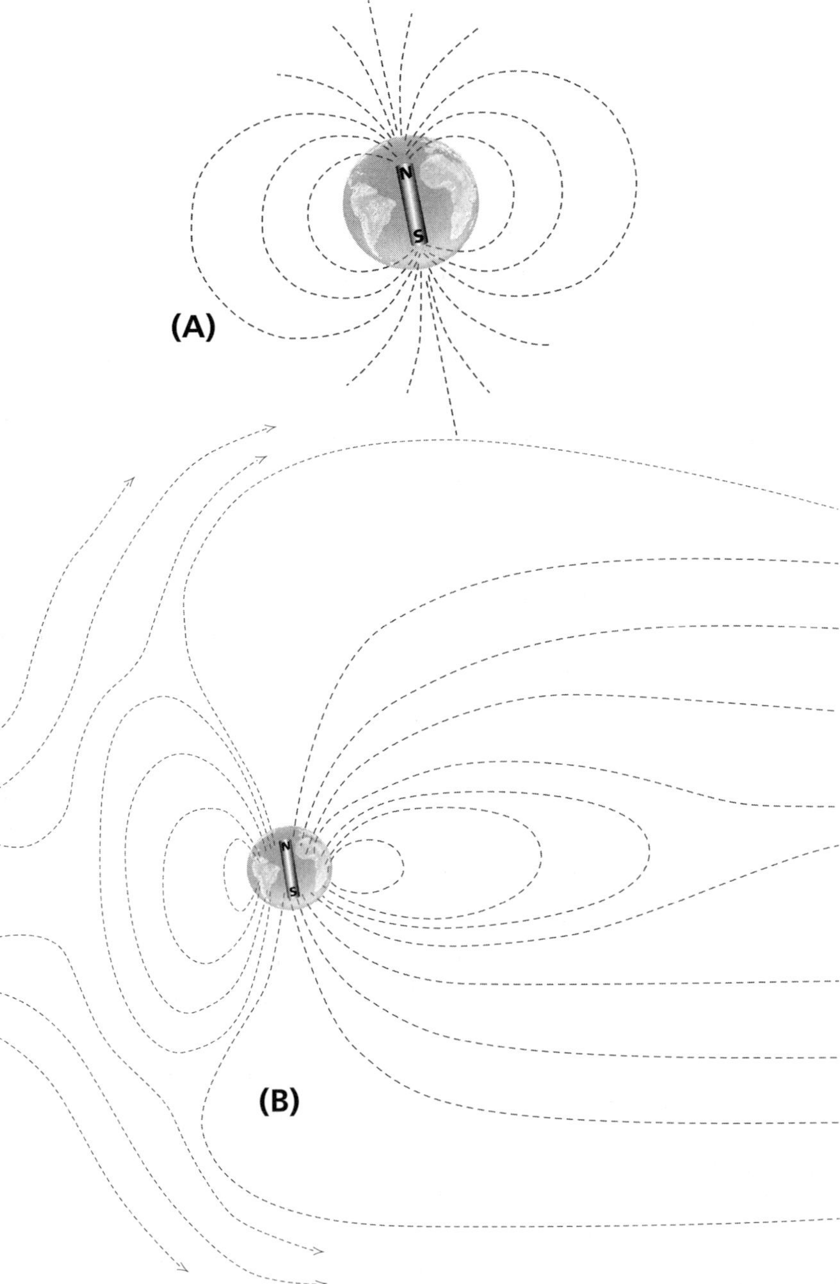

Figure 1. The earth's magnetic field lines are often idealized in illustrations as the same as a bar magnet with its north and south poles. (A) In reality, it is shaped by the electrical solar wind, which compresses the field lines (strengthening them) on the sunward side, and stretches the lines (weakening them) 'downwind' on the night side of our planet. (B) Therefore, as our spot on the earth's surface rotates toward the sun at dawn, the stretched geomagnetic field lines snap back, raising their strength. In areas with certain geological characteristics, this strengthening can be extremely dramatic and create powerful effects.

THE ELECTRIC HENGE

It is intriguing that ancient builders, as we shall see, repeatedly selected conductivity discontinuities for the sites of giant structures of earth and stone. Furthermore, most of these earth currents travel near the surface, in the uppermost three feet or so of ground. Cutting a three-foot deep ditch in the ground tends to block the flow.

A *henge* is a C-shaped ditch. Only a few also have stones. The ditches tend to have been dug a minimum of three feet deep. The open part of the 'C' is undisturbed ground that has not been cut by a ditch. Ground current, trying to flow across a henge, will be blocked by the ditch. Like water hitting a seawall, it will flow around the ditch, following the path of least resistance, which is the undisturbed ground in the middle of our 'C'. All the ground current will concentrate here in order to enter inside the area enclosed by the henge ditch. This is similar to what happens when a large tidal pond fills up and empties itself twice a day through a narrow opening. When the tide is changing, the current can be swift and surprisingly powerful within the narrow outlet.

Kaj and I have measured this effect on telluric current ourselves on site in England, as have others. However, we were quite surprised to see that Native Americans in the Midwest built similar henge-like structures.

MOUNDS AND PYRAMIDS: THE GREAT ATTRACTORS

The functioning of Native American mounds incorporates the physics of ground current as well as another feature well known to anyone who has seen a tall tree that has been struck by lightning. Any buildup of electric charge in the air will seek out the shortest course to connect with opposite electric charge in the ground. What is invisible to us is the mostly positive ground current that ends up concentrating as positive electric charge in the tree. Opposites attract, and the positive charge is drawn toward the predominantly negative electric charge in a thunderstorm. Those negatively charged electrons in the thunderhead likewise concentrate wherever an oppositely charged electric attractor is strongest, i.e., treetop. These two reach out to one another. In the process, they *ionize*, or electrify, some of the air between them. If this process proceeds far enough, a lightning bolt discharges and the tree is struck. What is important to bear in mind for our investigation is that not all such processes are energetic enough to be visible to the eye like a lightning strike.

If no thunderstorm is present, the location of the positive and negative charges is ordinarily just the opposite. Earth's atmosphere has a natural electric field that is generally positive (except during special local events like thunderstorms), while the surface itself is predominantly negative. The electric fields, like magnetic fields, possess field lines.

One of our instruments, the electrostatic voltmeter, can measure the amount of electric charge in the air. If you raise it from waist level to above your head on an average day, in an average place, you will see an increase of about fifty millivolts (mV). In certain special places and at certain times, it can rise a great deal more, as it did for us atop Tikal's Lost World pyramid.

Figure 2 shows how the atmosphere's electric field lines will concentrate at the top of a peak, and the negative charge of the ground will

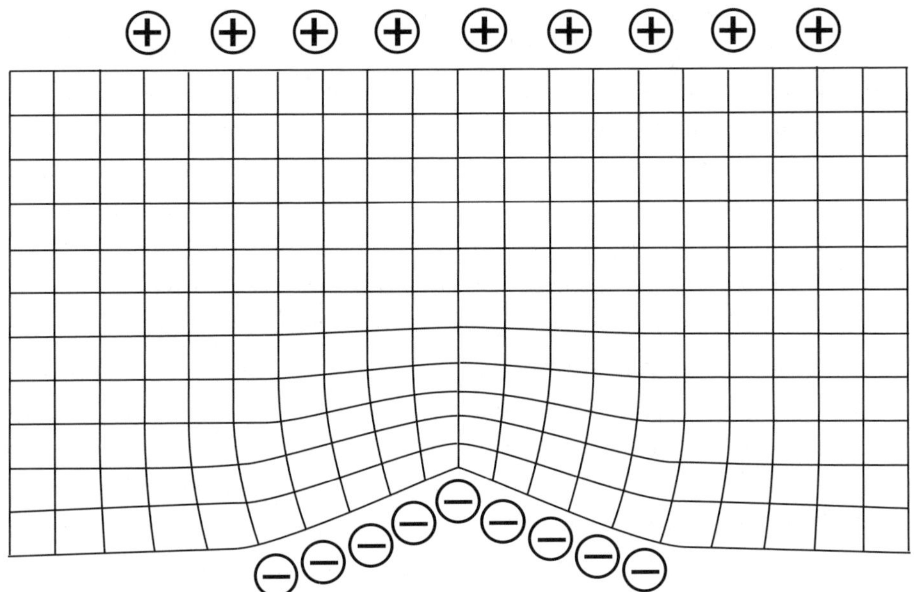

Figure 2. Any protrusion on the earth's surface concentrates the positive electric field lines of the atmosphere at the top. Likewise, negative charge from the entire ground area of the base is concentrated in the smaller area of the top. (Illustration after R. Reiter. *Relationships Between Atmospheric Electric Phenomena and Simultaneous Meteorological Conditions. I.* Air Research and Development Command, U.S. Air Force, 1960, by permission)

likewise concentrate at a peak. The trick for the ancient mound- and pyramid-builders (particularly in the lightning-rich Americas) was to build a mound on an electrically active spot like a conductivity discontinuity, then make the mound high enough and narrow enough to attract a dense bunching of atmospheric field lines – all without drawing a lightning strike. Neither these mounds nor the New World pyramids had pointed tops like in Egypt, lessening the chances of a lightning strike. However, many of the larger Native American earthen mounds had wooden temples on top, which every so many years would be struck by lightning and burned to the ground.

There are additional methods that the ancient architects used to concentrate and harness natural electromagnetic energies. For now, let us address the question of how these forces could possibly have improved agricultural results.

THE GENESIS OF A NEW THEORY

No one, to our knowledge, has ever reported a peculiar effect: that seeds placed at these structures and subjected to certain energies there will often grow dramatically better and give higher yields. In Guatemala, Kaj and I hoped to find evidence that would confirm our theory: that producing better crops was the motive for erecting these megalithic 'monuments.'

When we have explained this theory in person, we have inevitably been interrupted with the urgent question, "What in the world ever made you start thinking this way in the first place?" The answer involves the type of business that I was doing at the time. In 1993, I was helping to develop a new technology for treating seeds with electromagnetic energy. Prototype devices had improved seed performance dramatically by subjecting the seeds to a special type of carefully controlled shower of electrons. This treatment was not microwaves or irradiation, but something far gentler, more akin to static electricity (the type that makes a balloon stick to the ceiling when you rub it on your hair). It could dramatically change agricultural seeds when used very precisely, altering the physiology of the seeds and thereby the plants that grew from them.

The primary inventor of the electron shower seed treatment was Michigan biophysicist W.C. Levengood who had been working for over ten years on the electrophysiology of seeds. He had a few patents to his credit in the field, but this time, he had found something truly revolutionary. Seeds exposed to the right strengths of these gentle electron showers for the right amount of time showed drastic growth improvements. They germinated faster, grew through the seedling stage faster, matured faster in the field (as we were soon to find out), and were more resistant to stress of all kinds. Most importantly, they produced more food per acre. And all this productivity was accomplished without the use of a single chemical.

The improved resistance to stress was the most striking characteristic of all. Sweet corn seeds treated by these special showers and then planted in cold wet soil too early in the spring, were ten to twelve days more mature and an arm's length taller by midseason. Their ears of corn were ready for market sooner and there were more of them. The ears were larger and more uniform. We have since had commercial growers who used the process on sweet corn comment on the improved quality of the ears. Ninety percent of farmers, who have planted such improved seed in various countries, using a variety of crops, have returned each year to pay to have the process performed on their next year's seed. Growers who have tested a small amount have come back to insist that all their seed be treated next year or they will take their business elsewhere. Such stories include carrot growers in Canada, tomato farmers in China, and even whole provincial governments in China, where they are trying to grow soybeans on the same latitude as Nova Scotia, and finding the soybean plants can use all the help they can get. There the improved soybean seed emerges from the soil so much faster that it is apparent within days. The growth rate and final yields substantially exceed those of plants from non-treated seed. More examples of this phenomenon, from American universities and agricultural companies, can be seen on the company Website: www.proseedtech.com.

In later years, other test plantings were affected by extreme weather conditions and provided more visually dramatic evidence of the differences made by this energetic process. Carrots in any field that floods for several days early in the season usually produce a bumper crop of octopus-shaped roots that cannot be sold and are therefore called culls. When this happened in Ontario, Canada, in a field that was half planted with energized seed, we cut the cull rate in half. On the opposite side of the country, in the Pacific Northwest, a salt incursion in a carrot test plot a few days after planting killed forty percent of the carrot seedlings, except in the numerous plots sown with otherwise identical seed that we had treated. In Chinese provinces with acid soil, seeds treated with this energy produced more than twelve percent extra soybeans per acre and thirty percent more tomatoes, a vegetable that originated in pre-Incan Andes.

The process essentially subjects the seed to an electromagnetic impulse that prompts a natural response from the cell on the molecular level. It dramatically improves the plant's ability to withstand stress. While the biochemistry involved sounds technical, it is in fact the same response that occurs inside our own bodies when we jog or engage in other aerobic activity. The net result functions almost like a vaccination or inoculation against what is called *oxidative stress*. This is the primary cause of aging in our bodies and is the reason we take anti-oxidants like Vitamin E to guard our health. Oxidative stress is Mother Nature's biggest design flaw if you hope to live a long and vigorous life. Because of this insidious design flaw, we slowly lose vigor and start to break down a bit at a time. This is true for plants and animals. But we have found a way to tinker with this degeneration and improve plants' capabilities.

Oxygen is both our curse and our blessing. It is the high-test gasoline of metabolism but its 'exhaust,' so to speak, can be extremely toxic to those using it. Inside each of our cells are one to two hundred tiny energy factories called *mitochondria*. This is where the oxygen we breathe gets processed and used to make energy for our cells. The design flaw occurs because two to three percent of these oxygen molecules are processed improperly, and they emerge from the 'assembly line' in a damaged form. They are missing an electron. For reasons of stability, electrons around atoms and molecules occur in pairs. When you remove one, you then have a lone wolf electron that needs a partner. He will do his best to rip that partner out of anything he comes into contact with: cell wall membranes, DNA, etc. Those areas that have electrons stolen in this way now have their own problems and fail to function properly. In fact, the reason that rotten fruit gets soft is that a flood of free radicals are eating up the cell wall membranes inside.

When we run, we consume more oxygen. So more oxygen molecules are processed inside the mitochondria and that means that our steady two to three percent rate of mistakes produces greater numbers of free radicals. How can this be good for us? Well, strictly speaking, it isn't. What is good for us is the response of our cells to this threat. They make their own anti-oxidants, including Vitamin E, Vitamin C, and a dozen or more others that few of us have ever heard of. To simplify somewhat, these vitamins begin supplying that extra electron and thereby converting the dangerous free radical back to a normal, balanced oxygen molecule.

These natural anti-oxidants now gobble up the free radicals we produced during the run and then continue to hang around for a while. We now have more antioxidants and fewer free radicals than we did before the run, explaining how aerobic exercise extends our life. What's more, the next time we run and there is another spike in free radical production, our cells are better able to quickly produce more anti-oxidants than they were before they had this practice. This is why we compare these controlled free radical stresses to a vaccination, where the body is given a dose of something harmful in order to give the body's defenses practice at combating it, so that the next time they encounter the harm, they will be better able to deal with it. This process has been known in the plant world for a long time and is called 'stress hardening.' A specific form, 'cold hardening,' is familiar to many backyard gardeners.

In the spring when you take your tomato seedlings in their little cells out of the greenhouse to transplant them into the ground, you want to leave them first near the door of the greenhouse for a few days and let them gradually adjust to the colder conditions outside. Then you move them outside, but don't transplant them yet. During this period, the stress from the cold, like any environmental stress for plants, disrupts the oxygen processing and raises that two to three percent free radical rate to maybe five to six percent. Then the process we discussed above with increased anti-oxidant production occurs, and you now have a plant that can handle cold stress better next time. In fact, it can handle other stresses better too. Some researchers have noticed that cold hardening can also make a plant more resistant to disease or drought. What we have realized is that all stresses (heat, cold, drought, flood, disease, senescence) impact the plant the same way at the cellular level by worsening the free radical situation. One treatment with the above-mentioned electron shower can kick-start this process before the first seed hits the soil.

In this process, very low-energy electrons (running through the air at one millionth of an amp) coat the seed. As they are absorbed into the cell, they get absorbed as well into the mitochondria, where they are known to disrupt the oxygen processing in a way that increases the production of our free radicals. Then the cell's anti-oxidant defenses get involved, etc. In the end, you end up with a cell that can handle just about anything better.

The second worst design flaw of Mother Nature is that right where the free radicals emerge within the mitochondria resides the mitochondria's own DNA. Now, mitochondrial DNA is very unusual, not a double helix like normal DNA, because a billion or so years ago, mitochondria were separate, complete microorganisms. They were the world's first oxygen breathers. This made them Ferraris in a world of Volkswagens, and soon every Volkswagen wanted a Ferrari engine. Eventually the world's most important incident of symbiosis occurred, and a marriage was made between mitochondria and other microorganisms. They became one. Forever afterwards, mitochondria would live inside of cells that had a nucleus and the far more efficient double helix form for its DNA. That was when life could finally crawl out of the oceans, now that the poisonous atmosphere of toxic oxygen could be harnessed.

When some kind of damage (possibly a free radical) impacts a double helix, the odds are that what it takes out will not be very important, because most parts of the double helix do not hold genes. They seem to be there almost as filler, without a real role to play, which is why scientists named them junk DNA. Since junk DNA takes most of the hits, the double helix is fairly resistant to serious damage from small stresses. The mitochondrial DNA, on the other hand, is not a double helix but a simple ring, just like certain very ancient bacteria have today. It has very few 'junk' units, and the ring lies right next to where free radicals arise, so free radical damage is constantly hitting chunks of meaningful DNA. As you might imagine, once the DNA goes, things fall apart. And the DNA does go. The cell's natural anti-oxidant defenses are working all the time to defend against this threat, but they are always losing the battle. In the end, one bit at a time, cumulative damage occurs. This is why we have so much less energy as we age. A normal person aged fifty has half the number of mitochondria per cell as she or he did at age 20. Exercise can increase the number of mitochondria per cell in humans. In plants, the free radical equivalent of exercise, in carefully controlled doses, can do the same.

In our test plantings with seed companies and in our own fields, we repeatedly noticed an accelerated maturity. The plants were ready to be harvested earlier than seeds taken from the same seed lot, which did not receive an electron shower treatment. This early maturity allowed certain hybrids of corn, for example, to be planted further north than they usually could be grown and still attain full maturity.

While researching something else, I stumbled across a paper in a science journal that talked about a new school of thought that loosely calls itself 'mitochondrial inheritance.' It isn't inheritance in the sense of genes. It is simply this idea: As an embryo or a seed grows, its cells have to divide and duplicate themselves many times. It has long been known that when this happens, the double helix DNA in the cell's nucleus controls the operation. What has just been discovered more recently, is that the mitochondria still behave as if they were that separate organism that was co-opted by other cells so many eons ago. The mitochondria control their own division and replication. And now it is known that mitochondria with damaged DNA will make a damaged twin. Soon both of them will divide and replicate – making more damaged twins and so on. With each generation, more damaged copies are made, and the original damaged unit becomes an avalanche of damaged units. Ultimately, a child or an animal or a plant is born all right, but it bears less than perfect 'energy factories.'

For a plant, with limited amounts of light, food, and water, this process means that the plant's cells will be unable to make full use of the inputs. By contrast, a cell with fully intact mitochondrial DNA will make much more efficient use of the inputs, and, therefore, can complete its full life cycle on less input. In the real world, it matures early and, therefore, can be harvested before an average crop is ready. Once we understood this, we felt that we could tell everyone.

With our discovery of how electromagnetism can increase agricultural production, we could see what others had missed – the significance of heightened energy activity at ancient monuments.

CHAPTER 3

How Did They Know?

"Sir, I heare a report of a stonewall and strong fort in it, made all of stone, which is newly discovered at or neere Pequet [Gungywamp]. I should be glad to know the truth of it from your selfe, here being many strange reports about it."

– LETTER FROM JOHN PYNCHON TO JOHN WINTHROP, 30 NOVEMBER 1654.[1]

AT THIS POINT YOU MAY BE THINKING, "WELL, THIS IS AN INTERESTING theory, but these ancient peoples did not have electronic instruments. So how did they know about the presence of magnetic and electrical anomalies without the aid of magnetometers, electrostatic voltmeters, and ground electrodes?" The short answer is: They felt it.

In fact, the average person is often capable of detecting such minor magnetic differences – under the right conditions. A small percentage of people seem to be ultra-sensitive to magnetism and can locate anomalies more or less anytime. This is true for modern people, who live surrounded by the constant distraction of permanent magnetic anomalies in the steel beams of our buildings and cars, and who sleep and move in artificial electric fields. All of these elements must make us less sensitive than people who lived largely outdoors, without such desensitizing distractions. Yet, even we can feel electromagnetic changes at times. And shamans can almost always feel them. In this chapter, we will show you examples of this effect still occurring today.

Let us start with ourselves, regular people. Scientists in India, some twenty years ago, decided to measure just how much household electric fields might be affecting our minds and bodies.[2] They placed human volunteers, as well as rats, flat on their backs and passed a ring over their bodies, simulating the 110-volt alternating current fields in our homes, which oscillate fifty-five to sixty times per second. They measured blood pressure, pulse, brain waves, neurotransmitter levels in the blood, and also asked for comments from the humans. To their disappointment, they found no changes.

Now they decided to lower the rate of oscillation and see if that would produce physiological changes. It didn't, until they got all the way down to 0.01 oscillations per second. That is one oscillation per one hundred seconds! Then they saw fireworks. Three aspects of particular interest for our thesis are:

1. Oscillations of one per one hundred seconds are what frequently occur in the wee hours of the morning, when we get an occasional 'twang' of the magnetic field lines from disruptions in our planet's magnetosphere many miles above.

2. Oscillations of one per one hundred seconds are not what we call Alternating Current (AC), but are really more of a disturbance in Direct Current (DC). And DC is what is found in nature, not AC.

3. The strength of the magnetic field produced was fifty gammas or 50 nT, nanoteslas, equivalent to the strength of the earth's natural daily perturbations, only a fraction of the 300-500 gamma anomalies that we found at some mounds. (A *gamma* is a unit of magnetic field strength.)

Essentially, the human volunteers were far more susceptible to effects mimicking natural magnetic fields than they were to magnetic fields similar to those in their homes. Just what effects? Well, that depended on which way they were facing.

When their heads were pointing east, the rats failed to react but, humans reported a blissful reverie. With heads facing south or west, neither rat nor man reacted at all. But when heads were facing north and the staff triggered the magnetic fluctuations, everyone got very unhappy very fast. Humans reported anxiety, distress, panic, even nausea. The rats, who of course couldn't talk, finally weighed in on the self-reporting scale and screamed. Both species showed disruptions in every vital sign being monitored. And these sensitivities were found with average people, of no particular sensitivity.

The same strength in magnetic fluctuations (50nT or 50 gammas) has also been found to increase sudden unexpected death in epileptic rats. Such deaths among human epileptics peak during the wee hours of the morning, just as do the pre-dawn fluctuations in the geomagnetic field.[3]

NOT ALL DOWSING IS FALSE

Dowsing is traditionally the art of finding water, usually with the aid of a dowsing rod, which amplifies minute muscle contractions. The idea being that the presence of water causes the dowser to twitch. However, a broader definition of dowsing would include the ability to sense with the body a number of physical forces. Dowsers have long been the subject of admiration or disdain, depending on who you are talking to. Occasional scientific studies have pronounced that the performance of the subjects averaged no significant difference from what you would expect from random chance. However, there is one basic weakness in studying dowsers. You cannot just call up the Dowser's Union and ask them to send over a few journeyman dowsers, and please have them bring their certificates. All you can do is place an ad, asking for dowsers.

When your pool of self-described dowsers shows up, a large percentage will not actually be capable of the art. Some are people who obviously delude themselves. When you take their results and mix them in with people with a real ability, the percentage of accuracy is simply never going to reach the ninety five percent that science requires for statistical significance. Only then, when the odds are twenty to one that your findings are not random, is it accepted by the scientific community. Now, when a scientific report says 'no difference,' it often does not really mean no difference. In fact, the differences could be five to one against random chance, but since only the twenty to one 'gold standard' is accepted as statistical significance, the conclusion will often be worded 'no difference.'

Dr. Hans-Dieter Betz, at the University of Munich, is a little more imaginative than your run-of-the-mill scientist, because he thought about this problem and decided to simply analyze the data in a different way. Instead of lumping everyone's results together, he had forty-three subjects make multiple attempts and then analyzed the results of each individual for statistical significance. He and his team found that about two percent of the volunteers could hit the nail on the head almost every time, and scored better than ninety-five percent significance, or twenty-to-one odds against his performance being a matter of luck. For the best individual, it was 1,700 to one that his success was not just due to luck.[4,5]

Dr. Betz then took these individuals and subjected them to further testing. Some proved remarkably sensitive. When blindfolded and asked to walk the length of a board, they were supposed to point to various pieces of steel pipe that would be placed on the ground along the way. This was repeated again and again, as the objects were moved at random. Some fascinating insights to the nature of the dowsing response emerged.

First of all, it became clear that these gifted individuals were leading the rod rather than the other way around. A split second before the rod moved, their muscles would twitch. In other words, they didn't really need a rod. Secondly, some were capable of reliably detecting differences in magnetism of only a few gammas, or ten times more sensitive than the average people in the Indian experiment and more than sensitive enough to locate the much stronger anomalies that we have seen at megalithic sites. Unlike many of the volunteers of the Indian experiment, they were also consciously aware of the changes when they chose to be. Now, these results were all under laboratory conditions. Andrei Apostol wanted to know if the sensitives could perform in the field.

SENSITIVE TO EARTH MAGNETISM

Apostol is a Romanian physicist, who came to America from behind the former Iron Curtain to be able to indulge in a curiosity that others had called insatiable ever since he was a child. He knew that the first written reports of dowsing came from Germany in the 1500's, when miners used rods to locate underground bodies of metallic ores. (In Germany, as far back as 1747, such ores could also be located by looking for emissions of light from the ground.)[6]

In America, Apostol saw evidence similar to what we have found. He noticed how maps of magnetism and gravity again and again would show that ancient builders had singled out spots with anomalies in these forces and he, too, wanted an answer to the question, "How did they know?" So, when he heard about Dr. Betz's dowsing study, he immediately flew to Munich. He brought back one of the more talented individuals identified in the study, a dowser that he managed to intrigue with this enigma of the ancient Americans. In New York, he loaded this man, plus some very special equipment, into his station wagon and headed west. Apostol was not one to wait around for grant money; he just withdrew his savings and left town.

In the back of the station wagon, the sensitive would lie prone with a blindfold on, wired up to Apostol's apparatus that would measure any twitching of his arm muscles. Apostol would use a random number generator to choose from a selection of possible routes through an ancient site on a gravity anomaly. (Such anomalies were sometimes chosen for megalithic structures in the Americas.) The same method was used to select the time he began this route. In this way, the blindfolded volunteer wouldn't know where he was. But, as it turned out, when the car crossed areas where gravity changed force, his muscles began contracting and only relaxed as the car emerged out the other side, some twenty minutes later.

The dowser performed just as well standing up. Walking over the gravity anomaly of an underground cavern at Cave of the Mounds National Natural Landmark in Wisconsin, he produced the same pattern of muscle twitching. In other words, he could identify the location of a cave just by walking over a flat piece of ground above (Figure 3). Apostol mentions in his report that certain Native American groups would require an apprentice shaman to be able to locate a cave blindfolded, and perhaps they did it in a similar way. His work was considered solid enough to be published in the peer-reviewed *Journal of Scientific Exploration*,[7] and he was asked to present it at their annual conference in California. His scientific peers agreed that he was certainly onto something. We agreed, too. Several years later, we had cause to be particularly interested in his work, in connection with shamanism.

In recent years, several scientific studies have shown that humans are indeed sensitive to earth magnetism. Researchers have found traces of magnetite in humans, located in the sinuses of the ethmoid bone.[8,9] Such natural magnetic crystals are found in virtually all animals that use the earth's magnetic field to navigate; for instance, salmon, pigeons, and dolphins.[10]

ALTERING CONSCIOUSNESS

Dr. Michael Persinger is a Canadian psychologist who believes that magnetic fluctuations produced naturally by certain types of geology can have dramatic effects on the human mind. For fifteen years now, he

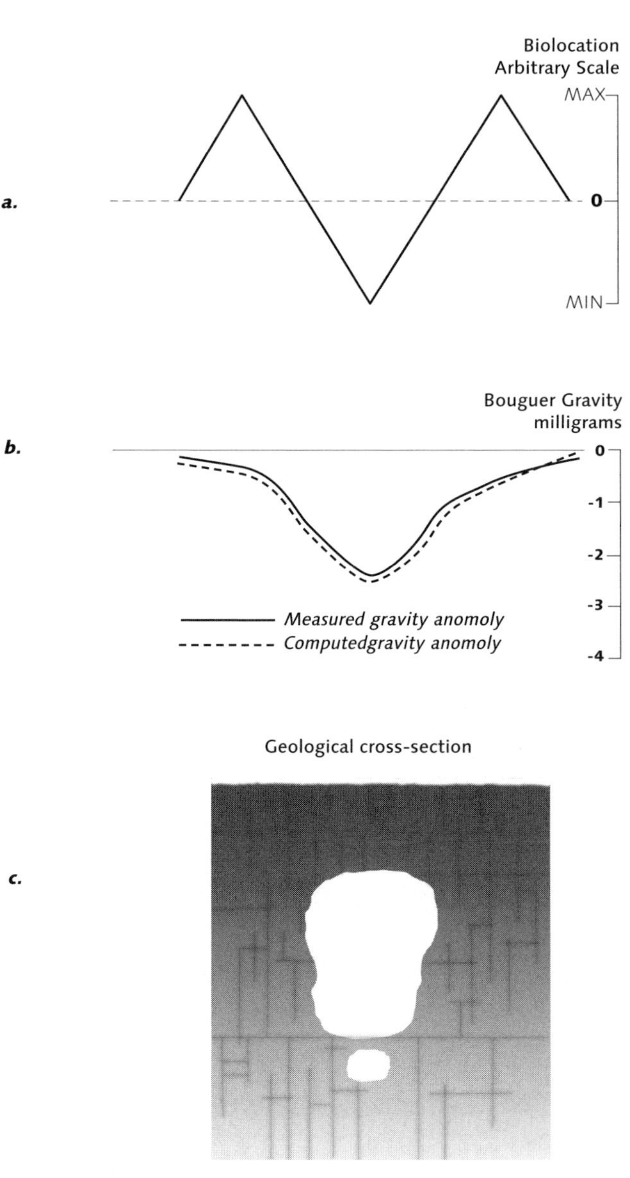

Figure 3. Andrei Apostol's sensitive volunteer showed muscle contractions (a), when he walked over a cave (c) at Cave of the Mounds National Natural Landmark, Wisconsin. The spot, naturally, had weaker gravity (b). Apostol relates that shamanic apprentices were required to be able to locate caves blindfolded. This is another example of how ancient peoples may have selected sites with unusual geophysical properties. (After illustration from Andrei Apostol. 'North American Indian Effigy Mounds: An Enigma at the Frontier of Archaeology and Geology.' *Journal of Scientific Exploration,* Vol. 9, No. 4, Article 3, 1996.)

has worked with 'the helmet,' as his volunteer subjects call it. It is an old football helmet wired so as to be able to produce fifty gamma magnetic fluctuations around the brain of a sitting subject. He chose this strength because it is consistent with the fluctuations often found in nature. His work has been publicized countless times in print and on television, because the results are so eerie.

Persinger reports that when he 'flips on the switch,' approximately thirty percent of volunteers report visionary-like experiences. These visions range from distortions of time and space to what the subjects call an encounter with the supernatural. Of course, much is determined by set and setting. If Persinger plays Gregorian chants in the background, a good percentage report having a religious experience. If he plays the theme notes from *Close Encounters of the Third Kind,* many report an experience similar to being abducted by a UFO.[11]

In our opinion, the giant outdoor laboratory that can confirm Persinger's findings in the real world lies just outside Albuquerque, New Mexico. Here, in Petroglyphs National Monument, the divide between two major zones of differing geology is provided by the eons-old course of the headwaters of the Rio Grande River. In this area, we have the largest conductivity discontinuity in the lower forty-eight states. We also have the biggest concentration of ancient Native American rock art in the U.S.

In the past decade and a half, the academic community has come to a consensus that most rock art was the work of shamans, illustrating their hallucinations during altered states of consciousness. This theory has been confirmed across the world in many other ancient cultures as well, and anyone interested in finding out more should explore the journal publications of Dr. David Lewis-Williams. Was the Persinger effect the connection between conductivity discontinuity and rock art? We decided to take our instruments there and find out.

From the moment I got out of the car and walked into the rocks that have tumbled off this basalt escarpment in Petroglyphs National Monument, readings on the electrostatic voltmeter showed that the rocks had electrical charge and were electrifying the air around them. Background control checks made it clear that this was not something happening in the whole area, just where the basalt outcropped, just where the pictures were carved. Late in the afternoon I had great luck. While measuring a distinctive outlying boulder with a carving of a shaman on it, the meter's needle began to rise until it was off scale. At first, only this rock seemed to affect the meter, but soon it became clear that there was, in fact, an enormous rise in electrification of the rock all along the escarpment. When the readings dropped a bit and the needle came back on scale, it was reading eight thousand volts per inch. This level of voltage is higher than we normally find in thunderstorms.

Running back toward the car, I made occasional stops for readings, showing that this disturbance had spread. Once in the car, I drove at seventy mph on the two-lane road, trying to get ahead of this spreading wave of electric charge. I raced a mile or so back up along the edge of the escarpment to the park's Visitor's Center and found that the disturbance had not hit there yet. So I waited. Within ten minutes it showed. As I drove along the several-mile length of the park rocks, I was able to follow the spread of the anomalous charge. It took about forty-five minutes to travel three miles. Usually, electric current travels faster than that. But it was positively charged; not the lightning-fast electrons of what we usually think of as electrical current, but rather the heavier positive ions, which are slower moving. Whatever it was – it was extremely powerful.

When I returned to the Visitor's Center for more readings, the rangers asked me what I was doing. They knew that this place was the nation's biggest conductivity discontinuity, and one of them told us something interesting. With an amused smile, she said, "You know, sometimes I get these New Age types in here who tell me they just like to go up and sit in the rocks and 'feel the energy.' I always thought they were a bunch of flakes, but you're telling me there just might be something to it."

GUNGYWAMP SURPRISES

Eastern American residents, seeking a taste of these forces, might consider a visit to Gungywamp Swamp in southeastern Connecticut. Here, high ground in a still uninhabited marshland is sprinkled with magnetic anomalies, and rock chambers are built over many of them. One chamber has a special shaft aligned so that the rays of the rising sun on the winter solstice penetrate the heart of the chamber and illuminate a connecting miniature chamber, only three feet high. Visiting these sites – as well as the infamous Cliff of Tears – we received some surprises along the way.

The first surprise came during our walk down the trail behind the late Dave Barron, a laconic, white-haired Connecticut Yankee with a wickedly dry sense of humor. He had been president of the Gungywamp Society for what seemed like forever, at least to Dave. "I made the classic mistake in a non-profit foundation," he told us. "I didn't show up the day we selected a president, so they picked me, because I was not there to defend myself." Of course, his work was so exceptional that no one ever wanted anyone else.

"You know, I've been taking people on tours of this place for twenty-five years, and I have heard some making these crazy claims about energies and 'vibes,'" he said, pointing an index finger at his temple and making little circles. "Lots of them brought dowsing rods, and I thought, hey, let them have their little fun; it's harmless. But over the years, I started to notice people getting the same reactions at the same spots, over and over again. So I decided to give it a try."

Barron broke out his own dowsing rods: cut-off pieces of a wire hanger, bent and stuck loosely in plastic sockets, which he held at chest height. After a few minutes, they started spinning fast enough to produce a high-pitched whine, with the metal rods rubbing the edge of the plastic sockets as they twirled. "See what I mean?" Barron smiled. "This is a pretty active day."

Walking behind him, we looked at our electrostatic voltmeter, and every time we got to a spot where his rods twirled, our readings would spike. On one stretch of the trail, beside a row of standing stone slabs, his twirling and our meter peaked repeatedly at half a dozen spots, located equidistant to one another. At each spot, there seemed to be an electrified vertical curtain of air, about a foot thick, stretching clear across the trail. We didn't believe it at first, so we double checked and triple checked. The columns of electrical air rose from the ground to as high as we could reach. "See what I mean?" Barron said again. "On any active day, the rods go crazy here."

What we were really hunting here, however, was not the dowsing response, or even the rock chambers. We wanted to visit The Cliff of Tears, about which we had heard so much. It was thus named, because a large number of people had simply broken down here and started sobbing uncontrollably – for no apparent reason.

But Barron was a retired speech therapist and a no-nonsense New Englander. "I didn't put much stock in this stuff when I started giving tours.

But then I began to witness something pretty funny. One day, I took a bus from Alcoholics Anonymous on a walking tour here, and *four* of them got nose bleeds, right at The Cliff of Tears! One or two I would have chalked up to coincidence, but *four* made me wonder. Then women I know and trust started telling me about spontaneously beginning to menstruate while standing there! So, we did a study. Nurses measured twenty volunteers that day. The one thing that changed to a degree of statistical significance was blood pressure."

Walking on, Barron continued his banter, occasionally pointing out a particularly strong dowsing response. So we paid no special attention when at one point on this woodland trail, he casually gestured into the underbrush and, seemingly off the cuff, told us, "We usually get an interesting little energy line here, running up the hill."

He kept walking, but when our voltmeter detected the line, our curiosity led us to follow it through unbroken underbrush, up a small hill. Our progress was halted by a rock wall at the top, but about ten feet out from the base of this rock was a six-foot-wide circle, where the voltmeter reading was so strong and abrupt that it caused us to take the Lord's name in vain. Behind us there was chuckling, evolving into outright laughter, from the other members of the Gungywamp Society who had come with us. This uproar was followed by Barron's dry voice, "You're standing at The Cliff of Tears."

The electrostatic readings inside this six-foot circle remained steady, so we turned our attention and our magnetometer on the cliff face itself. As we got close to it, the readings began climbing at an unexpected clip. Startled, we wondered what could be causing such a surge, until we saw – running horizontally across the face of the cliff, just above head height – a six-inch wide brown band of magnetite ore.

WEST COAST POWER POINTS

Residents of California do not have far to go in order to experience some of the energies discussed here. A remarkable and long-established 'New Age' center called the Ojai Foundation sits on a plateau of some power. Significantly, the Ojai Valley is the only place in California where the San Andreas fault runs east-west. Since geomagnetic field lines run north-south, the lion's share of the telluric ground currents they produce can be expected to do the same. This theory makes an east-west section of this part of the San Andreas fault a powerful conductivity discontinuity. This has been confirmed by our own readings.

Ed Sherwood is an Englishman of unusual talents, married, and settled in Los Angeles with his wife Kris. They did not know about the geology of Ojai, or about our instrument surveys there when they selected it as the place to spend their wedding anniversary each year. All they needed to know was that Ed could see light balls there, sometimes sporadic, but often prevalent. These are the same sort of light balls that emerged in our photos from Tikal (Plates 3-8). Ed has a library of light-ball photos, but the difference with him is that for years he has been able to feel their presence. He has been known to announce to a group of friends outdoors at night, "They're here!" and begin snapping away. The light balls will show up in the photos, often shoulder to shoulder. When he announces, "They're gone now," his photos back him up. And when he pronounces, "Now they've moved over there," again his photos confirm his statements. In recent years, Ed has been able to train himself to see the light balls as well. He claims that he can train most people to see them, too, if they are capable of total concentration and can spare a few days.

He first contacted us, and sent us photos, after discussing the light balls with the director of Ojai, who showed him our report. This report relates how on two occasions we measured the kind of rapid early morning alterations of geomagnetism seen near conductivity discontinuities. One of these visits was full of wonder.

On our first morning there, sunrise had technically occurred but not at Ojai. An odd fog pocketed this miniature sloping plateau. It was like pea soup, completely obscured the sun, and had these odd concentric rings extending above the rest of the fog bank in the direction of the sun. We took magnetometer readings at the small stone circle made by Ojai visitors near the top of the plateau. Readings were quite uniform until the sun broke through the fog and then, suddenly, sky-rocketed four hundred gammas in fifteen minutes. As far as we know, this is not supposed to happen. It is known that telluric earth currents will jump when the sun breaks through a cloud or fog, allowing the fair weather electric field of the atmosphere to connect with the earth. These electric currents will produce changes in the local magnetic field, which will be exaggerated at a conductivity discontinuity. But in this case, the normal morning change in the geomagnetic field was completely blocked until the fog bank dissipated. Then the field jumped to where it should be and remained stable the rest of the day. On the second, clear day, the normal geomagnetic field changes happened gradually. But this dense fog, while not quite violating the laws of physics, certainly stretched them. Therefore, we were tickled to learn that this spot had been labeled 'The Dragon's Eye' by a prior visiting group of Tibetan monks who chose it as their spot for morning meditations. They told the staff, "This is where the power is."

The monks also indicated that the ridge stretching beyond the circle was high powered and the afternoon of that first day, we saw why. About an hour before sunset, while we were walking along the ridge-top trail, readings on the electrostatic voltmeter began to soar. Even on the high setting, the needle danced back and forth in abrupt swings until it finally climbed off scale positive and stayed there. We can count on one hand the number of times this has happened to us, so it is no surprise that sensitives like Ed Sherwood, the Tibetan monks, and thousands of others continue to return to the Ojai Foundation for mind-altering experiences of a natural order.

SEDONA ENERGIES

It had not been our intention to go to Sedona, Arizona, until people, hearing about our readings elsewhere, implored us to go there. They exclaimed, "Oh, how can you *not* have taken your instruments to Sedona!? You go to an energy vortex there and, boy, you'll get readings all right, because you can *feel* the energy!" So we went, not really expecting to find much, but, once again, learned the importance of an open mind.

Sedona certainly is staggeringly beautiful, surrounded by striking red rock cliffs and spires. Control readings miles before we reached Sedona showed us what we had expected: that all this red rock country is strongly magnetic. The red is, after all, from oxidized iron in the sandstone. This usually means that most, but not all, of the iron will be non-oxidized magnetite. Readings of 570 gammas in the vertical axis were a good five gammas more than we got before hitting 'red rock country.' After dining on buffalo steak, we drove out the dirt roads in the hills beyond to watch the magnetic readings drop. And drop they did. Readings fell hard and fast after sunset.

Within hours, we had seen enough Kokopelli figures to last us a lifetime. We also got our fill of the phrase 'energy vortex.' Nevertheless, the next morning we dutifully obtained a map that located and gave directions to the most popular energy vortices. They were sandstone buttes at one end of town, vertical rock spires perched atop a humped base hundreds of feet high, like images out of a John Ford western in Monument Valley. They turned out to be centered on a negative magnetic anomaly (a spot with lower geomagnetic field strength) and lie at a gradient (or border zone) of two areas of differing magnetic field strength. This was the classic geomagnetic profile we had come to expect from earth mounds or rock chambers. The geomagnetic characteristics of the two vortices we visited here reminded us of what we had found in other spots that had natural electromagnetic energies powerful enough for Native Americans to have selected them for special purposes. With a nod to Kokopelli, we drove off into a red rock sunset.

WONDERS OF BEAR BUTTE

It has become accepted wisdom in the anthropological community in recent years that ancient rock art was created in trance, probably by shamans who were depicting their hallucinations. As we have seen, the largest amount of rock art in the U.S. is located on a geological structure that creates powerful electrical ground currents at Petroglyphs National Monument. These geological structures will themselves create changing magnetic fields, which in turn can alter consciousness.

Shamans sought out these energies to alter their consciousness, something that still happens today among modern Native Americans. The Black Hills of South Dakota have long been considered sacred ground by the Sioux, as well as by the tribes living there before them. The Black Hills are also one of the nation's leading conductivity discontinuities – and again Kaj and I wondered if there was a connection. There is.

For us, it started with a personal experience on Harney Peak, at over seven-thousand feet, the highest point of the Black Hills and long considered the most sacred peak. We had been measuring electrical ground current two-hundred feet below the summit and needed a break. I leaned back with closed eyes, resting my head on the cliff face. Dizziness and disorientation were the immediate results. Raising my head out of contact with the rock stopped the effect, while contact would initiate it all over again. Getting up to inspect, we found that the rock in question was a vertical vein of quartz running through the cliff face. Our meter showed that it had a strong electrical charge and was also charging the air next to it. We surveyed the entire cliff face and found that this was only the case in two small spots, both of which were veins of quartz. This effect made perfect sense, because quartz stores electric charge like no other mineral, which is why it is used in watches.

Now we began to understand why the Sioux insisted in their treaties that no other reservation ground would do, they had to keep the Black Hills, whether or not the white man wanted them for gold. The following day we were to stumble into a tradition that is still very much alive today.

At the Black Hills Visitor's Center, you can look down on a marvelous three-dimensional table map of the region. When you are wondering what might be an interesting site to visit, one stands out from the rest. The site, a lone spike in a flat land, lies to the north. If you travel south from Canada over one thousand miles of flat ground, this would be the first vertical break you encounter. Bear Butte is a geological twin of Devil's Tower in Wyoming, featured in *Close Encounters of the Third Kind.* Both are volcanic plugs, highly metallic lava that cooled and hardened eons ago in the throat of volcanoes, which have themselves since eroded. But unlike its twin to the west, Bear Butte is still covered by dirt and looks simply like a tall, very steep hill with a razor-edged flat top.

Today, this mountain is a state park with a trail. The signs read: "Open to visitors during daylight hours only. All white visitors are banned from the park after dusk." Then at the foot of the trail: "Please stay on the path at all times. Respect native traditions. Do not touch any objects off the trail. These rules are seriously enforced. You are visitors here. These grounds are still actively used for religious purposes." By the time we started up the trail, we noted two items of importance. Tied to the scattered trees were brightly colored pieces of cotton cloth, called prayer flags by Native Americans. And secondly, the place was littered with magnetic anomalies, running the gamut from one to four-hundred gammas, similar to what we have found on mounds. Wherever there was a magnetic anomaly strong enough for our magnetometer to detect, the spot had been marked by a particularly dense cluster of prayer flags. Occasionally, a medicine bundle would hang suspended from a tree.

A stiff twenty-five-knot breeze was blowing out of the north, and the prayer flags stood straight out, fluttering furiously and providing the only sound in this otherwise silent and deserted world. At the densest cloth cluster of all, twin-festooned trees framed a stunning panorama of the plains to the north, at the edge of a thousand-foot cliff containing the most powerful magnetic anomaly, an incredible 900 gammas, we have ever found (Plate 9). Leaning precariously over the abyss with the probe, we realized that this two-foot-wide anomaly on a projecting rock spire was exactly consistent with how lightning strikes will magnetize rocks. Nearby was an altar-shaped stone that contained a quartz crystal and a clam shell. The stone was marked with the names of visitors like Lindsey Walking Eagle, Rick Thundershield, Crystal Highwolf, and Lillian Whirlwind Horse.

We continued, trudging up the steep trail, pulled irresistibly now by this eerie, compelling place. The top of Bear Butte is a flat ridge, about five to ten yards wide and maybe one hundred yards long. From here, you look almost straight down on the roofs of the tiny dots of cars below. The top contains five striking magnetic anomalies, detectable on our magnetometer. They had also been detected by someone else, because every single one of them had been prominently marked. In four of the five places, a shallow pit had been dug, each encircled by a low wall of rocks.

The fifth site was unique. No hole, no wall, but a sleeping bag stashed in a tree nearby. This, at five hundred gammas, was the strongest of the five ridge-top anomalies. Instead of a circle, it was surrounded by a square, five feet on one side. In the corners stood small piles of rocks, holding two-foot willow poles that marked the cardinal directions: a red cotton flag in the north, a yellow in the east, a white in the south, and a black in the west. Connecting the four flags and forming the sides of the square were strings covered with hundreds of knotted bits of brightly colored cloth (Plate 10). In the center of the square stood two erect Y-shaped willow sticks, holding up a long, horizontal willow stick. It was apparent that the shaman had chosen the area with the strongest magnetic activity.

VISION QUEST

In *Yuwipi. Vision and Experience in Oglala Ritual,*[12] a contemporary Oglala Sioux from the Pine Ridge Reservation describes a vision quest he had in the 1960s. He had asked his spiritual leader for help in his unsuccessful struggle against alcoholism. The following week, his spiritual leader drove him to the foot of a butte in the Black Hills and walked him to the top. Here he was seated in an enclosure, the same as the one we saw atop Bear Butte, surrounded by an array of colored flags, knotted strings, and sticks. Identical flags marked the same directions that we had recorded. The sticks were all cut from a sacred willow that morning. It turned out that the horizontal willow stick held up by the two vertical, forked pieces was used to hold the medicine bundle of the shaman. It was left with the narrator who stayed alone for the night. There had been no extensive fasting, and no drugs were involved.

He sat here through the night, lost in personal thought. Near dawn (the time of the strongest geomagnetic fluctuations), his thoughts turned into visions. The first one was auditory, a drumming that swelled to such magnitude that he feared it would split his eardrums, until he realized the source of the deafening beats. As he looked down at the ground between his legs he could see ants, their feet touching the ground with each step, and he realized that they were the drummers, creating this roar. Next he noticed a thundercloud moving right at the butte until, at the last moment, it split in two, each half sweeping by him on either side. As this happened, faces materialized in the cloud wall and leaned out, stretching their necks as they loomed over him and screamed. He wrote about these events fifteen years later and had never had another drink.

Persinger reported that the effect of geomagnetic fluctuations in humans is probably mediated by the pineal gland, which is most sensitive to the fluctuations late at night. The fluctuations are linked with increasing frequency of episodes of epilepsy, called the 'sacred disease' by the ancients because of its disproportionate occurrence in shamans and oracles.[13] Such natural fluctuations are also linked with certain hallucinations.[14]

We also find it interesting to note that Persinger's work has found the religious experiences associated with magnetic field fluctuations to take place in the brain's temporal lobe.[15] This region of the brain lies next to the temples. It is worth recalling the Olmec giant basalt heads that all had the north magnetic pole located at the temple by the carvers.

The tradition of vision quest, which we had observed on Bear Butte, had been developed by the childhood friend and spiritual mentor of Crazy Horse, who led Crazy Horse to Bear Butte. A Sioux shaman, he taught the system of the colored flags and willow sticks to others, still identical in detail now 150 years later. The meadow by the parking lot here, loosely decorated with cotton 'flags,' is where the Sioux would gather in the morning to wait for Crazy Horse to descend from the butte and preach his visions (Plate 11). In Pine Ridge today, there are shamans who can pick out magnetic anomalies and use them to produce visions in their people. The National Park Service now allows Native Americans to continue ritual use of Devil's Tower – Bear Butte's geological twin.

OF MUSHROOMS AND FLYING SHAMANS

There is yet another way that the ancient builders may have been able to identify power spots, namely the physical effects that these spots produce on plant growth. Jiri Dvorak is a Czech hydrogeologist. His company uses scientific-sensing equipment to spot the types of geological structures where municipalities can most profitably drill wells for water. These are the structures where water rises to the surface and, therefore, where natural springs occur most often. They are *interfluves:* the boundary between two different levels of aquifer. In his long career of prospecting throughout Europe, Dvorak has found that these interfluves are by far the most prolific spots for hunting mushrooms.[16] We knew from our research with electricity and plant growth that mushroom growth is dramatically amplified by even very low-level electric currents in the ground.

As Gordon Wasson has amply demonstrated in the classic *Flesh of the Gods,*[17] one sacred ritual of ancient, pre-Christian Europe centered around the conical mushroom *Amanita muscaria*. Its powerful, mind-altering properties were so prized by shamans that it was hunted to extinction on the Indian subcontinent (where yoga then was invented to replace it). It is hard to exaggerate the importance of this fungus to Western consciousness. One set of Amanita users who, in the early twentieth century, still practiced a late Ice Age lifestyle were the Lapps, or Sami, of far northern Scandinavia and Russia. They were semi-nomadic reindeer herders, living in yurts and driving about in reindeer-drawn sleighs.[18] Their shamans wore pointed red hats, symbolizing the mushroom that was their sacrament. The shamans also wore red capes, symbolizing their flight through the air when they consumed the mushroom on their most sacred night of the year – the Winter Solstice (then December 25). So from their mushroom hunting, the Eurasian shamans, the repository of knowledge in all pre-literate societies, would have known all about the growth-enhancing powers of these same types of geology that were selected by the megalithic builders.

Starr Fuentes trained for six years with contemporary Mayan shamans in Meso-America. When she first heard of our hypothesis, she asked us, "So, do you think the ancient builders were following the ley lines?" Ley lines, as described by Alfred Watkins, are a series of straight lines connecting ancient sacred sites. We told her no, because the ley lines we have seen on maps are too straight to be produced by any geological structures. But Starr explained, "I don't mean those kinds of ley lines. I mean the ones that affect plants." She went on to explain how Mayan shamans know that to improve a plant's health you can take it to a 'ley line,' leave it for a period of hours to days (depending on the desired effect), and then bring it back home. We asked her how to find these ley lines, and she replied, "You just follow the clay."[19] Starr has no training in geology, and at that point did not know about our research regarding interfluves. But these energy-producing interfluves are typically associated with surface deposits of clay.

Thus, we have first-hand reporting from two continents about the observed connection between improved plant growth and the same geological phenomena that were harnessed by the megalithic builders. These builders, therefore, would have had two means of discovering appropriate locations: sensing and observation.

CHAPTER 4

Return to the Lost World

"Maize lay undiscovered in a mountain cave underneath a large rock, until a bolt of lightning penetrated the cave and split the rock apart, revealing the seed."

— POPOL VUH, *SACRED BOOK OF THE QUICHE MAYA*

TIKAL, FIRST CITY OF THE MAYAN CIVILIZATION, IS THE LINK BETWEEN OLMEC culture and Mayan. It, therefore, seemed to us to be the right place in Meso-America to take our instruments. We knew about the connection between prosperity and mounds in Olmec civilization. We also knew from England (described in Chapter Nine) that some man-made Bronze Age mounds were highly active, electromagnetically speaking. So it made sense for us to visit Tikal.

Once we arrived, we did what all first-time visitors do, go straight to the Grand Plaza, a World Heritage site and, incidentally, the spot depicted as the rebel base in the first *Star Wars* movie. Surely, we thought, this is where the power must reside. So when our guide Luis knocked on our door at the Jaguar Inn, just before 3 AM, we dutifully followed him into the pre-dawn darkness. Prior experience had taught us that this was the right time of the day to catch the geomagnetic fluctuations and telluric ground currents that we were stalking.

Atop what Coe and Diehl had dubbed 'Temple II' – but what everyone else calls the King's Pyramid – our small team set up. We used a fluxgate magnetometer to catch the geomagnetic changes, ground electrodes and a hundred feet of wire to measure the telluric currents, and two electrostatic voltmeters to measure electric charge in the air. We sat down and waited. Few things can try your patience more than staring at a meter hour after hour. Computers were not feasible alternatives as laptop batteries would not give us the amount of time we needed. We recorded magnetic changes and electric charge in the air, but nothing unusual – nothing we couldn't find in the jungle far from the ruins, nothing we couldn't find at home, for that matter (Plates 3 and 7). The pyramid that we had just 'staked out' was one of the last ones built here. Around 600 AD, the people of Tikal were conquered by a northern tribe. After a century of subjugation, a rebel chief led a successful revolt, putting the locals back into power. In celebration of his triumph, two fantastic stepped pyramids were built, today called the King's and the Queen's Pyramids. The former is dead as a doornail, electrically speaking, and the Queen's has been off limits since a tourist, descending the steep staircase, tumbled down and broke his neck. Most of the hundreds of other structures in this metropolis had been built before these two pyramids. So, we reasoned, the engineers of the early buildings would have had their pick of locations. If there was energetic ground around here, it would have been theirs for the grabbing. The fact that a later, purely political pyramid did not exhibit electrical charge suggested to us that the Olmec's knowledge was possibly lost in the cultural collapse of the seventh century. We decided to test our theory.

PULSE OF THE PYRAMID

Set several hundred yards away from the Grand Plaza and its teeming visitors lies the first structure in this first city of the Maya. It has been named the Lost World Pyramid because of its isolated position. Erected around 600 BC, more than a thousand years before the more elaborate structures, it is much humbler than most of the newer pyramids, lower and flat on top. During the incredible 1,300 years that it had been in use, it was expanded six times. So we set our sights here.

On the first morning, we found impressive electrical and magnetic charges on top of this hundred-foot pyramid. Elated as we were, our excitement this day was dwarfed by what we found on our next to last morning in Tikal, where this book began. Over seven hours that day, our array of scientific instruments recorded an electrical dance of sorts between the atmosphere and our planet's surface, a dance in which this 2,600-year-old structure acted as intermediary. The amount of electric charge in the air was greater than what we had usually found during thunderstorms. Fortunately, we had obtained the assistance of the U.S.

Geological Survey in Colorado, who had advised us in our purchase and use of ground electrodes. These electrodes showed that the ground of this tiny plateau was pulsing with the strongest electric currents we had found anywhere: up to six hundred millivolts/kilometer. This reading is enormous for telluric currents. Intriguingly, it is in the same range as those measured on Piton de la Fournaise. So perhaps the Olmec, who designed the first version of this pyramid, were in fact trying to duplicate something that happened naturally in their volcanic homeland, just as they seemed to be at San Lorenzo. However, measurements of current in the ground of the plateau surrounding the pyramid showed that, while the whole miniature plateau was pulsing, it seemed to be the ground beneath the pyramid that was producing most of the action. Geologists use this method to identify certain subsurface geological features that are known to conduct current the best. It certainly looks as though the Lost World Pyramid was placed directly atop one of these. As we shall see in Chapter Ten, the Americas are probably not the only place where this was done.

The top of Tikal's Lost World Pyramid (Plate 8) was clearly concentrating this ground current and linking up with an airborne electric field that ranged from 1,100 volts on the ground to 1,720 volts when we lifted a voltmeter over our heads to eight feet. Airborne readings will normally rise with height because the ground and the atmosphere tend to carry opposite charges. But from ground level to eight feet usually shows a difference of perhaps ninety volts or so (which is what had happened on the King's Pyramid). Here we had a significant difference: 540 volts averaged over eight readings on different days. One day, these readings showed an average change of 908 volts from ground to eight feet up, or ten times that of the gradient on the King's Pyramid.

These voltages might sound lethal, and if it were household current, they could be. However, static electric charge in the air is a different type of electricity and even a thousand volts is not dangerous. In fact, you can generate a stronger charge by rubbing your hair with a rubber balloon. That's why our hair did not stand on end atop the pyramid. Nevertheless, these readings are way out of the ordinary and can have some profound biological effects, as we are about to see.

Furthermore, the electric charge in the air jumped dramatically at the time of official sunrise, for about forty minutes or so. This is due to the change in the earth's magnetic field at sunrise. But in three mornings on the Lost World Pyramid, it averaged 827 volts vs. only 186 volts on the King's Pyramid. Since such exaggerations of the dawn surge are known to be associated with certain geological configurations, it looked like the oldest pyramid was built on just that sort of ground, while the later ones were not. If the knowledge that had led to the building of the early pyramids was lost when Tikal's rulers were overthrown, then the practical benefits would have also been lost. The electrically dead King's Pyramid had no effect on seed. It is interesting to note that not long after this possible cultural decay, the civilization of Mayan Tikal crashed.

Readings at the Lost World Pyramid were always higher on the four corners. Even below the top, holding a meter to a corner always gave 45-60 volts/inch stronger readings than any other part. This was in keeping with what science knows to be the behavior of electric charge. As mentioned earlier, geophysicists have made measurements atop Mexican volcanoes, showing that electric charge concentrates at sharp edges. Perhaps this effect was copied by the Olmec from their volcanic homeland, and perhaps they shared this knowledge with the Maya in building this very pyramid on which we now stood. As shown in Figure 2, all these effects are in keeping with how electric ground charge is known to interact with electric atmospheric charge at a hilltop. To get control readings for that morning, I hurried to the tallest pyramid in Tikal, Temple IV, a later structure that was built for political prestige, and where tourists today gather to watch the sunrise. None of the electrical forces were present here. Only the early Lost World Pyramid was pulsing with energy.

Building a pyramid has two complementary effects: It concentrates any electric charge in the ground at the top, and it bunches up the atmosphere's electric field lines at the top. Figure 2 also shows why you would think hard about whether or not to give your pyramid a pointed top or a flat one. If you have as much energy as seemed to be present at the Lost World plateau, and you concentrate it too much by using a pointed top (like an Egyptian pyramid), then in a rainforest environment with plenty of lightning strikes during thunderstorms, the amount of energy concentrated at the top may be too much for your purpose. The main god of the Aztec, the Mayans' cultural descendants, was Tlaloc, 'He who makes things grow.' He was also god of rain and lightning,[1] and the largest Aztec pyramid, on a scale with Egypt's, was dedicated to him.

MAIZE ON 'MAIZE MOUNTAINS'

Back on the Lost World Pyramid, we laid out the most sensitive of all our monitors: corn seed, or maize. Interestingly, the Mayans called their pyramids 'maize mountains.'[2] They were often placed over or beside 'fertility caves.' In fact, long before Mayan civilization, the Olmec began the tradition of bringing seed to altars in caves, decorated with fertility symbols, afterwards taking the seed home with them. From years in Asia, we knew that the Chinese, among others, do not waste such food offerings by throwing them away; they eat them. Leaving the offering for a day on an altar to the traditional earth god, for example, they then remove and consume it. Perhaps the Mayans took it home and planted it.

Anthropologists of the nineteenth century reported common offerings of maize seed in Mayan caves with jaguar fertility figurines.[3] Today, white-robed Mayans can be seen emerging from the jungle before dawn to perform ceremonies atop the pyramids of Tikal. Part of the ritual involves placing corn seed and beans on the highest stairs. After the gods have had time to 'take the essence' from the offering, the Mayans take it to their home to be used.[4] They always miss some, and we found dozens of seeds scattered widely across the top of the King's Pyramid.

One can easily imagine how all of this might have begun. If Mayans originally had caves that they used for rituals, perhaps visions quests, some of these caves might have become shrines where offerings were made. Every traditional culture uses offerings to the gods or nature forces that they revere. Food is the most common offering, and for Mayans the basic food staple has always been corn. If someone took corn seed home after it had been left for a while in a cave as an offering, planted it, and saw dramatically better growth, it would not take long for word to spread. Soon the cave would be thought of as a fertility cave. Then the Olmec introduction of pyramids could have been married to this tradition and we have the pyramids placed over or next to fertility caves, just what the ethnographers have reported.

As described in later chapters, we had previously placed seed on Native American earthen mounds in central U.S., as well as in pre-Columbian rock chambers in New England – which look like artificial caves – with fascinating results. Here, at Tikal, we brought dried local corn seed, purchased a few days before in a Mayan hill town. This local corn seed was not high quality, which was precisely what we wanted: seed whose vigor was as close as possible to the ancient stocks that were in use before modern breeding and seed handling.

We spread the samples on the flat rock atop the pyramid and waited, still taking readings on our instruments (Plate 11). While the howler monkeys roared and birds started calling, the early morning fog drifted eerily through the treetops level with our vantage point, dawn returning colors to the world.

LIGHT BALLS AND IMPROVED GROWTH

At 11 AM, we packed up and left, exhausted and hungry, but elated. We sensed we had struck gold. Little did we realize that the best results were yet to come. Months later, back in the U.S., the corn seed from the Lost World Pyramid was germinated at Pinelandia Biophysical Laboratory in Michigan by biophysicist W.C. Levengood.

The photos of these germinated seeds revealed some remarkable results. The Mayan corn seed that we spread out on the electrically inactive King's Pyramid (Plate 13) did no better than the control seed kept in our hotel room (Plate 12). In fact, they did a bit worse – probably the influence of the heavy dew that settled on them that morning. But growth of the seed placed the first morning on the electrically active Lost World Pyramid was clearly improved (Plate 14). And, finally, seed left atop the Lost World Pyramid on the day of highest electrical activity was drastically improved (Plate 15).

Furthermore, our Tikal photos were bursting with visible confirmation of these electromagnetic forces. On several prior occasions, we had noticed that balls of light appeared in flash photos taken at ancient sites. As Dr. Levengood explained it, the photons of the flash add energy to the already energetic molecules of the electrified air. These molecules absorb this extra energy, driving them to a still higher energy state. Within a fraction of a second, however, the molecules drop back down to their original energy state, radiating the extra flash energy as light. All this happens too fast for the eye to catch, but not for the camera. Since we had recorded the actual times of exposure for our photos as well as the times of all instrument readings, we could confirm that the photos were choked with overlapping light balls when the air was highly electrified (Plates 4 and 5). As the amount of electric charge dropped, so did the density of these light balls in photos taken at the same vistas (Plates 6 and 8).

All the factors that we went to Guatemala to look for had now come together and had been recorded. We were on our way to linking a series of startling findings into a coherent picture.

CHAPTER 5

Before the Inca

"To create, out of the human spirit, something which did not exist before."

— William Faulkner, Nobel Prize speech

In the eastern part of the mountain range Cordillera Blanca in northern Peru, in a valley about 10,300 feet above sea level and near the present-day village of Chavin de Huantar, lies the ruins of what was the first dazzling cultural empire of the South American cradle of agriculture. What forces gave birth to this civilization that flourished here, long before the Inca?

Some ten miles from the village, the mountain Nevado de Huantsan – believed to be the abode of the local guardian deities – soars to 20,976 feet. Even today, the residents of Chavin consider the nearby peaks to be inhabited by spirits. From Nevado de Huantsan, a stream called Wacheksa tumbles down to join the Mosna River. Something about this small river was unusual – it became a sacred site for oracle[1], a process that involves the alteration of consciousness. People visiting the stream felt that they could communicate with the powers of the Earth here. This practice continued here for some 2,400 years, until the Spanish conquest.

Shortly after 900 BC, a stone structure was erected at the site. This building was not simple, nor was it small, and since no one lived nearby, great effort was required to erect it. The entire valley could only have supported around 2,000 people, so many from outside the region came to help erect the large stone building, half the size of a football field and forty-five feet high. It had few doors and was honeycombed with passages, stairways, air vents, and windowless rooms with large, flat stones for ceilings, stones that were used nowhere else in the area.[2]

Essential to what happened here seems to have been the oracular stream. Throughout the lower level, the water was channeled through stone passages down the natural slope, the stone courses fanning out below the building, but not leading to any fields or houses. Their purpose was not to irrigate.[3] It was something else. Could it be energy? Certainly, the similar Olmec structures of San Lorenzo come to mind when we wonder about the purpose of the ancient Peruvian builders.

Shortly after the building at Chavin de Huantar was erected, visitors began streaming in from every direction. Despite its remote location, Chavin grew to an enormous complex, twenty times its original size.[4] Beside the stone building, a sunken courtyard was dug and its sides lined with stone. The canals angled off to also channel the rushing water under the court. As we shall see later, such sunken courtyards were common in Andean architecture. It seems that the electrically active rushing water was the central force of this semi-subterranean feature. And Chavin de Huantar prospered. Before long, gold and silver began showing up in this under-populated area.[5]

All these features, however, would later find their embodiment on a far larger scale in the south of Peru, with the most fascinating civilization ever to grace the continent. It too flourished long before the Inca.

TIWANAKU OF TITICACA

The highlands of Peru and Bolivia, the *altiplano,* is a harsh place. Fierce winds blow over this 13,000 feet high, treeless plateau, where a merciless sun burns by day and temperatures drop below freezing at night. People here graze herds of llama and alpaca, but the thought of farming in this climate, on this poor, thin soil seems ludicrous, and today its agricultural base supports only a widely scattered population.

The altiplano is dotted with lakes, some of which are remarkably beautiful, colored bright green or red by trillions of minute algae. Queen among these lakes is Titicaca – the highest navigable lake in the world. A thousand years before the Inca, a people lived on the southern shores of this enormous lake, in the village of Tiwanaku (or Tihuanaco). Their society was dominated by the principles of duality and complementarity of the sexes, shown for example in their carved images of paired males and females.[6]

You would expect these bleak highlands to be the last place for a high civilization to develop. But nearly two-thousand years ago, Tiwanaku began to stand out from its neighbors. Why? Its land was no better, and it boasted no desirable harbor on the lake. It did not lie on lucrative trade routes, nor sit astride valuable mineral deposits. It seems that none of the factors, traditionally regarded by historians as central to the emergence of power and wealth were present. Yet, at its height, circa 500 AD, Tiwanaku teemed with a population of about 40,000, probably making it the largest city outside Asia in its day. Yet food production – mysteriously – seemed to be no problem, and the town was renowned for its agricultural surpluses. Even the poorest peasants grew an estimated three times the food they required and traded the excess, becoming prosperous.[7]

One key to the success of Tiwanaku's food production was raised-bed agriculture. This method was also used with great success by the Maya and the Aztec. During the dry season, the lake bottom was dug up and heaped into ridges that would be above the wet season water level. These ridges were flattened into tiny farm plots, becoming surrounded by water during the rainy season. Mud scraped from the watery bottom each year was dumped on the plot as fertilizer. The high water table would keep the soil of the platform relatively moist, while the warm air above the canals around the raised plots tended to mitigate the damaging effects of temperature extremes.

Modern studies, recreating raised-bed farms on Lake Titicaca have shown that with this technique, potatoes were produced at triple the rate of modern methods.[8] Yet, the raised-bed method has not been widely re-adopted today because construction of the beds requires immense, sustained efforts. The people of Tiwanaku were clearly energetic enough to make such efforts. Aerial views of the southern shores of Titicaca show seemingly endless stretches of the remains of ancient raised beds, farmed for centuries when the water level of the lake was higher.

However important raised-bed farming was to Tiwanaku at its height, this was not what originally distinguished the village from its neighbors. Raised beds did not appear in great numbers here until 500-600 AD, centuries after the town had begun to prosper.[9]

After eighteen years of excavating and studying Tiwanaku, archaeologist Alan Kolata of the University of Chicago concluded that what gave the village an advantage over other nearby villages was the fact that after its people had commenced erecting huge megalithic structures, Tiwanaku began to prosper in an unprecedented way. People from the surrounding regions were streaming into the village on what is normally referred to as 'religious pilgrimages.'

Yet Kolata emphasizes that even the present-day religion of the Aymara (direct descendants of the Tiwanakans) – despite its long intermarriage with Catholicism – focuses only on practical results. Kolata notes, "On the high plateau, spiritual insight flows from hard living on the land, from the yearly cycle of planting, irrigating, weeding, and harvesting. Ritual is grounded in reality, and reality becomes ritual. Through their spiritual life, the Aymara seek health, abundance, and fertility, not a vague sense of harmony with Mother Earth. They are, in a real sense, mystics, but not of the esoteric kind enshrined in our common imagination. Here the spiritual dimension is found in the ordinary acts and objects of everyday life: in the smell of eucalyptus burning in the hearth, in the rich, smoky flavor of a new potato baked in a clay oven carved from the earth."[10]

ARTIFICIAL HILLS AND MEGALITHS

One thing, however, distinguished Tiwanaku from the other villages – its geology. The village sits in a small valley, flanked by seismically active faults. Up through the cracks from the major faults had welled rich deposits of basalt.

The type of basalt, cherished by the Tiwanakans for their megalith building is a cousin of diorite, the material of the famous bluestones of Stonehenge. Like diorite, this rock is also magnetic – and so much associated with the volcanic Andes that geologists have named it *andesite*.[11] None of the andesite used in the monumental buildings was quarried near the village. In fact, the nearest major quarries seem to have been forty-five miles away across the lake on the far side of the twin peninsulas of Copocabana and Huata, where small deposits of andesite are found near the water. Amazingly, it is assumed that the Tiwanakans floated quarried blocks across the lake from here because the unthinkable alternative was to haul them over mountains. And if you think the builders of Stonehenge were impressive in hauling four-ton blocks of diorite, try to imagine the people of Tiwanaku transporting andesite monoliths weighing up to 140 tons.

Near the village, they erected buildings of andesite and red sandstone, the most magnetic type. Some of the buildings were constructed on top of an artificial hill, called Akapana, fifty-six feet high (Figure 4). Early European explorers just assumed Akapana was a natural hill. Even when told otherwise by the Aymara, they refused to believe that anyone had created this massive structure in such a forlorn place.

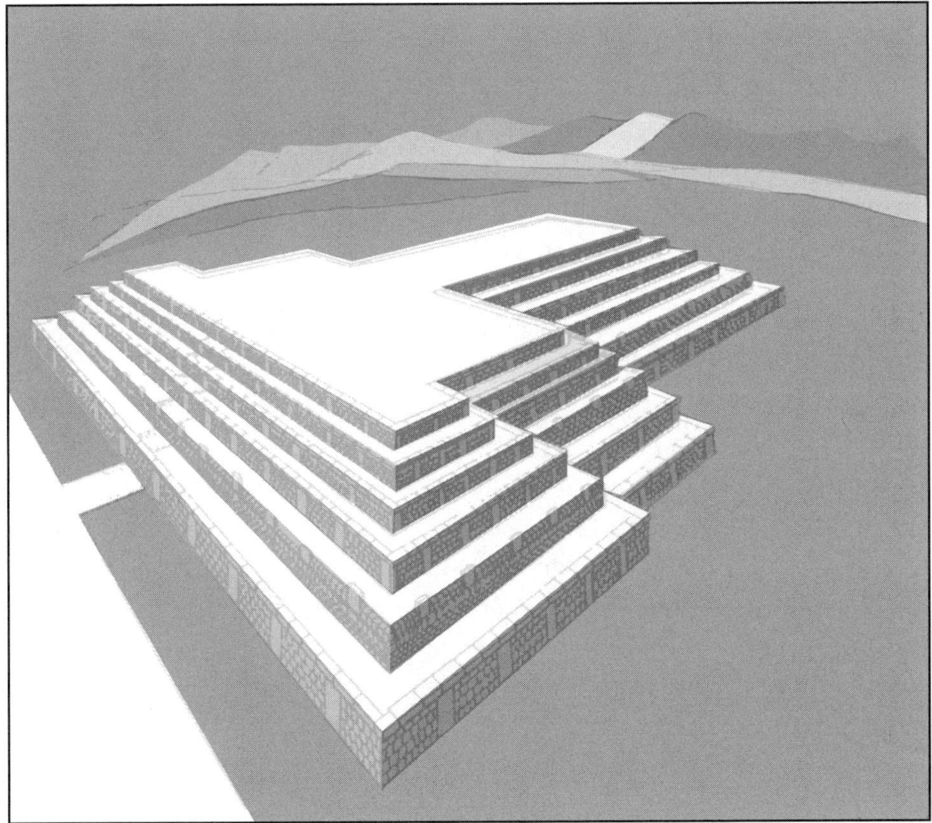

Figure 4. Centuries before the Inca empire, people of Tiwanaku moved enormous stones across Lake Titicaca to create massive structures, including the Akapana Pyramid shown here. All that archaeologists found at the top were utilitarian-looking rooms, shaped much like the rock chambers of New England. Inside were only the remains of corn seed and small potatoes – perhaps seed potatoes? (After Alan L. Kolata. *Valley of the Spirits: A Journey into the Lost Realm of the Aymara.* John Wiley & Sons, New York, 1996.)

The body of the pyramid was composed of andesite blocks, fitted so tightly together that, even today, a playing card cannot be inserted between them (Plate 16). In similar cases in England, human skulls, long bones, and mandibles (lower jaws) were part of dedicatory burials, accompanying the laying of the foundation.[12] Then clay, with all its electrical conductivity, was piled up, followed by a layer of a special type of gravel from the lake. These pebbles were a rich green in color, no doubt because of their high content of copper, the most electrically conductive mineral besides pure gold.[13] Clay went atop the copper pebbles, followed by more pebbles and again clay.

ALTERED CONSCIOUSNESS

Even today, the andesite outcrop high above the ruined city of Tiwanaku is sacred to local Aymaras who make pilgrimages up there every spring. Ancient petroglyphs carved into the rock walls here show that the site was special to the original inhabitants as well. Archaeologist Kolata accompanied a group of Ayamara shamans on a pilgrimage and states that his consciousness became dramatically altered as soon as they set foot on a field of andesite bedrock, underlain by a rushing stream.

Kolata tells us, "I know at this point of transition I felt distinctly strange, possessed of an exquisite, yet disturbing, hypersensitivity. My senses seemed abnormally acute. I could focus with great clarity on individual sounds and sights, but my overall perceptions of people and landscapes seemed indistinct. Even now I remember small details of objects, of places, of scenes with extraordinary precision, while my impression of the whole experience is inchoate . . . If this stage of the pilgrimage was evoking these emotional states in me, what was happening to the mind and sensibilities of my Aymara companions, and especially to the shamans?"[14]

The shamans continued to the far edge of the andesite, to a point where it meets a sandstone ridge and a granite outcrop where a spring forced its way out of the rock wall. Sitting here, as sunset was followed by darkness, the chief shaman underwent an epileptic-like convulsion and entered a trance. During this trance, the shaman would sense whether or not this year's growing season would be a good one. How could he know? We wonder whether the electrical lability of the shaman's temporal lobes enabled him to gauge the degree of electric energy in the earth. A spot like this should be electromagnetically powerful and, as we have seen, even the archaeologist was dramatically affected.

The shamans attempted to mediate with the *ispallas* of their crops. According to one Aymara shaman, the "spirits of the ispallas are the spirits of the fields, the spirits of the harvest, the forces within the seeds of the edible plants." These *ispallas* are reported to be the same forces that can draw a shaman into trance.[15] *Ispa* means both 'twin' and a large potato, their main crop, thought to be struck by lightning and then called 'potato with two faces.'[16] The *ispallas*, lightning, twins, shamans, and the proper growth of food crops are all entangled in the mind of the Aymara.[17] All but the twins are recurring factors in our investigation.

DOWN THE DRAIN

Throughout the interior of the Akapana pyramid, special drains made of andesite and sandstone were laid out.[18] They would alternate between closed channels inside the hill and open channels outside, weaving in and out, plunging abruptly to the next level in small waterfalls and finally cascading to the andesite foundation five stories below (Plate 16). Some commentators have described these spillways as intended to imitate mountain streams, running now inside, now outside rock faces. The drains were almost hermetically sealed so that very little of the electrically charged air could escape – just as they had been in Chavin, and in the Olmec area. The splashing of the water running downward inside the drains would generate electrical charge. Like the Olmec, the Tiwanakans might have noticed the electrical properties of running water, as well as the effect on the human mind of magnetic forces. Also, the effects on their crops might have been noticed, and they wanted to copy this effect.

The wet and highly conductive rock segments of the drain were made all the more electrically conductive by being bound together with clamps made of copper. Into T-shaped grooves, carved into the edges of adjacent stone blocks, liquid copper was poured, binding the blocks tightly together and forming an electrically conductive bridge between them[19] (Plate 16). The entire length of the drains, coursing throughout the inside of the hill, would become electrified. The top layer consisted of copper gravel. In the center, an artificial pond held tens of thousands of gallons of water that could be released into the drain system by lifting a sluice gate, again similar to the Olmec system.

According to archaeologist Arthur Posnansky, the stone buildings atop Akapana were made of andesite.[20] They were too cramped, plain, and unadorned to have ever been designed for ceremony and ritual. It is unlikely that anybody would be living there with no hearths for the frigid climate and no kitchen. All that was found inside these windowless buildings was broken shards of undecorated, utilitarian pottery and the remains of corn and small potatoes, perhaps seed corn and seed potatoes.

Even today, seed potatoes are a common ritual offering at Aymara shrines. If the Tiwanakans ultimately removed and used their offerings (like the Maya), this could be the way they had discovered the seed enhancement potential of the early shrines.

Around 600 AD, the drains of Akapana got clogged and stopped flowing. Another manmade hill was erected less than a mile away at Puma Punku even closer to the fault. It was threaded within by similar andesite drains. Here, they not only ran down through the man-made mound to the base of the artificially modified rock outcrop on which the complex was erected, but also ran through farm fields. As one commentator put it: "thus unambiguously linking the summit complex with agricultural productivity."[21] In fact, it is generally assumed that the leaders of Tiwanaku, as with the later Inca, derived their power and legitimacy from the fact that they guaranteed agricultural and reproductive success.[22]

SEED ENHANCEMENT?

A semi-subterranean courtyard, called Kalasaya, was several hundred yards on one side, again lined with andesite and red sandstone. This huge rectangular enclosure seems to have originally been marked off by standing pillars of natural, unworked *magnetite* and andesite. This pattern is reminiscent of similar spaces marked off by pillars of a different type of basalt, *serpentine*, at the Olmec site of La Venta. The fact that the inside surface of the pillars – and only the inside surface – was smoothed and flattened is reminiscent of Stonehenge. Later, the Tiwanakans further enclosed the Kalasaya by filling the spaces between the pillars with blocks of cut stone, as if to more completely contain something within its walls.

Sunrise on the two Equinoxes bisects the courtyard, but none of the design was actually necessary to that purpose. The whole structure functions to achieve an energetic end. The drain from Akapana ran around the inner edge at the base of the stones, possibly to help fill the courtyard with airborne ions. Flanking the courtyard is a series of small stone rooms. They have always puzzled investigators, having no windows, benches, or decoration – no indication that they were ever used for ceremonial as well as standard utilitarian purposes. They had large, flat stone slab ceilings and were designed with small doors that would minimize air turbulence inside, just what you would need to separate electrically charged components of the air and generate pulses. The builders went to the extra trouble of providing some with sliding doors of heavy stone slabs to seal and open them at will.

As we shall later discover in Chapter Six, slab-roofed rock chambers in New England dramatically enhanced the growth of wheat, bean, and corn seed. Is this what the builders of Tiwanaku were after? Would it justify such Herculean labors? On the altiplano, during the frigid high-altitude nights, one of the gravest threats to farmers is cold damage to their seedlings early in the year. Today, on the altiplano, frost heavily damages crops on average three years out of five and entirely destroys one crop in five years. Getting the crop through the vulnerable seedling stage as quickly as possible would lessen this threat. This is exactly what the New England rock chambers have done to seed placed in them: increased the rate of germination, improved seedling growth and seedling tolerance to cold and stress, and, finally, made such seed produce up to three times more food. Remember also, it has been estimated that the peasants of Tiwanaku grew three times the amount produced by other Andean peasants.

On the northeast tip of Copocabana Peninsula, on a hill above Lake Titicaca, lies a cluster of dry-wall, corbel-vaulted, oval rock chambers, named Poto Poto. When they were excavated, many contained no bones and therefore were never the tombs that archaeologists label them as. Others contained bones, but they were 'intrusive' burials – placed here long after the structure was built. The chambers have large, flat rock roof slabs similar to those of all the chambers known to improve seed performance. Another cluster of both rectangular and beehive-shaped rock chambers lies on Taquiri Island, close to the northeast corner of Copocabana Peninsula. One of the chambers held twelve skulls,[23] reminiscent of the interred skulls at the causewayed enclosures that we shall later visit in England. All these chambers look across a few miles of the lake to Tiwanaku.

THE ROCK THAT GAVE BIRTH TO THE SUN

The center of the mythology and religion of the Tiwanakans – as well as their Inca successors – was an object on the island that is today called Isla del Sol ('Island of the Sun'). The Inca have often been called sun worshippers, but their own Temple of the Sun stands second to this object of reverence on the still sacred Isla del Sol. The island was considered the center of the world and site of the origin of life, not only by the people of Tiwanaku, but even by the Inca a thousand years later.

Thichicasa (and Titicaca) comes from *Titi-kala,* meaning 'Rock of the Wild Cat,'[24] named after the supreme mythological figure of the jaguar, which throughout the Americas was considered the symbol of fertility. From this rock, the island, and the enormous lake itself take their names. Legend says that when the whole world was dark, the sun itself was born from this rock.[25] Could this legend have arisen from the presence of a ball of orange light rising off this rock one night? Similar balls have appeared around the world near fault lines and conductivity discontinuities. It is known that after the Inca kicked the Tiwanakans out, the chief Inca would travel all the way here from Cuzco for each New Year's dawn and, by his presence, help to usher life back into the world.

Whatever might have made it famous, the Rock of the Cat is certainly not revered for its vista. Sited on the northwest neck of the island, the view from here is described as 'monotonous,' whereas elsewhere on the island there are spectacular views of the highest range of the snowcapped Andes, towering over the deep blue waters of the lake. In a saddle between bumps on a ridge, the Rock of the Cat appears as a shapeless lump, measuring 130 by 190 feet. The Rock is red sandstone, the most magnetic type.

According to legend, there was a sacred cave in the rock, but in fact, the 'cave' is a natural recess with a simple altar in it. On either side of the recess is a two-meter-sized nodule of exposed *illmenite*, a magnetic mineral. These nodules are also considered sacred. The electromagnetic forces of these rocks may be enough to alter people's consciousness. On the ground below are what archaeologist Adolph Bandelier called finely worked slabs of 'prismatic' andesite, brought from outside the island.[26] The 'prismatic' part of his description reminds us of the seemingly crystallized surface of the highly magnetic roof slabs in the rock chamber shown in Plate 17. Like the stream at Chavin and the tunnel below the Aztec Pyramid of the Sun, the Rock of the Cat was said to have oracular effects.

Another phenomenon here might also have garnered notice. In 1910, Bandelier was told that no birds ever fly over the rock.[27] One wonders if it interferes with their magnetic navigation senses.

THE PLACE WHERE PEOPLE LOSE THEMSELVES

The eastern half of Isla del Sol is primarily comprised of a 200-meter-thick layer of dark limestone, extending both above and below the brilliant blue of the lake surface.[28] Riddled with springs, it rises in cliffs and steep hillsides from the lake. As we shall see in Chapter Nine, the electromagnetic properties of such carbonaceous strata, like chalk and limestone, are powerful when water moves through them.

On this limestone, the Tiwanakans concentrated their building. The structures, ranging up to half the size of a football field, were made of stone, and were filled with mazes of tiny rooms. There were no windows, no chimneys, no light, usually only one small door, and no appreciable air circulation. As dwellings, they would be torture chambers.

The ceilings are interesting, generally being large stone slabs, just like in the rock chambers or dolmens. Some are highly corbelled to assume a beehive-like shape inside and can be expected to create the kind of air separation present in the seed enhancing chambers of New England. Some of the larger chambers had andesite doorways. Whatever these structures were used for, they were not tombs.[29] The Tiwanakans must have been satisfied with the results, because they continued to erect bigger buildings. A thousand years later, the Incas threw the Tiwanakans off the island, claimed possession of it for themselves, and built even bigger structures.

Interestingly, the buildings invariably follow the lay of the land. If a boulder stuck out of the hillside, it was not removed but incorporated into a wall like the rock chamber shown in Plate 19. Although the hillsides all over the island were terraced for farming, no one ever changed the ground level in these buildings. They simply followed the natural contours, although that meant irregularities in the building everywhere. Such behavior was puzzling, unless there was something valuable in the ground that the Tiwanakans did not want to alter.

Four hundred and fifty feet away from The Rock of the Cat is the building that in the beginning of the seventeenth century was known as Chincana, or "the place where people lose themselves."[30] Too cramped and unadorned to be a palace – as it is colloquially referred to today in travel guides – and too big to be a warehouse for the limited population of the island, it consists of twenty rooms with large stone slabs on their corbelled, vaulted ceilings. It was built to surround a spring. The walls are two to eight feet thick but only eight feet high – similar to other rock chambers around the world. Bandelier states, "The whole complex has but one small air-hole, to which the name of window cannot in justice be given."[31]

We shall later discuss how geomagnetic fluctuations are maximized by what geophysicists call the 'island' and 'peninsula' effect. The Rock of the Cat lies on the northwestern neck of the island. This entire side of the island is composed primarily of Devonian shale and quartzite. What is fascinating is that the major seismic fault in the area runs right along this side of the island. It hugs the coast and runs up along the promontory upon which The Rock sits. This combination should subject the area to the well-known piezo-electric effect in which quartz under pressure, such as tectonic strain, develops electric charge. It is believed to be this mechanism that causes production of earthquake lights, mysterious glowing balls, and even large hemispheres of corona discharge during earthquakes. They might even look like a sun rising.

The entire island is a perfect site for maximizing natural electromagnetic forces, and the visual appearance of these forces may be enshrined in the local folklore. The sloping point of the Sicuyu promontory, where such forces could be expected to be at their maximum, is covered with offertory cysts, from which gold statues have been recovered. When the conquistadors began pillaging this site, the local inhabitants would make their offerings by lowering them onto the underwater extension of the ridge. Even today, no one crosses over this ridge in a boat, unless to make an offering, the purpose of which is to seek fertility for the crops.[32]

END OF THE GOLDEN AGE

According to Kolata, who is arguably the premier authority today on this site, the Tiwanakans developed huge agricultural surpluses, attained in the most unlikely of places, and used them as the basis for trade, transforming the village into the center of the most powerful kingdom in South America until the Inca.[33] While the empire of the Inca lasted only 150 years, the Tiwanakans reigned for more than a thousand. Only a devastating, centuries-long drought could end it, around 1200 AD.[34]

In what they considered superior wisdom, the Spanish conquerors instructed the locals what to grow. Wheat, olives, and other European crops were introduced, crops totally unsuitable for the high-altitude climate of the Andes and the altiplano. The transformation was performed with what the Spanish considered modern farming practices – simply meaning the European ones. These crops failed, and the formerly rich cultures plummeted into poverty.

And still today it is so. The current sad state of Andean agriculture is the inevitable result of a paradox of modern farming. The methods employed are those espoused by twenty-first century farmers, using heavy doses of artificial additives to their fields, such as chemical fertilizer, herbicides, and pesticides. Today's Peruvian and Bolivian farmers cannot afford these additives.

The result is exhausted land. The nutrients of the Andean and the altiplano soils are long depleted. The pathetic returns from farming this land can only support a fraction of the number of people who inhabited the Tiwanakan and Incan empires. A new approach is needed desperately. Or perhaps, it should be an old approach.

CHAPTER 6

Rock Chambers of New England

"Rules of Reasoning in Philosophy, Rule II (. . .) We are certainly not to relinquish the evidence of experiments for the sake of dreams and vain fictions of our own devising; nor are we to recede from the analogy of Nature, which is want to be simple and always consonant to itself."

— SIR ISAAC NEWTON, *THE SYSTEM OF THE WORLD*

NORTH AMERICANS TEND TO THINK OF MEGALITHIC STRUCTURES AS something you have to go overseas to experience. The truth is that two-thirds of the U.S. population lives within a few hours drive of one. Just one hour north of New York City lie some exceptional examples, and we are grateful to the scientific investigators who first introduced us to the energetic nature of these structures.

Dr. Bruce Cornet is a geologist whose palaeontological discoveries were hailed by *Discover Magazine* as one of the top hundred science stories in the late nineties. In his spare time, he and Connecticut science teacher Phil Imbrogno studied reports of unusual lights in the area. Cornet and Imbrogno worked in the Hudson River Valley, New York, charting repeated sightings of mysterious lights moving through the sky. Mapping the movements of these lights over time, they found that their trajectories would often trace back to common points of origin. When they mapped these points and compared them to a detailed magnetic map of the area, they found that the lights would always cluster near a negative magnetic anomaly – a spot where the earth's magnetic field is weaker than in the area around it. "And," says Imbrogno, "when we visited these sites, we invariably found one of these rock chambers right at the anomaly."[1] What he is referring to are the mysterious, pre-Columbian rock chambers found all over New England, presumed to have been built by Native American tribes.

Cornet was able to get access to a state-of-the-art proton precession magnetometer from the Lamont-Doherty Geophysical Observatory of Columbia University. When used at the rock chambers, it would always show a spot with a negative magnetic anomaly in the ground – right outside the single doorway to the chamber (Figure 5).

As Imbrogno and Cornet questioned local residents, they began to hear interesting tales. A man walking his dog around eleven o'clock one night heard a deep, low hum emanating from the direction of a rock chamber, "like the sound an electrical transformer makes." As he came around a bend, into sight of the chamber entrance, he saw that it was dimly illuminated from within by a blood-red glow. Other stories entail ghostlike entities inside of or emerging from rock chambers.[2]

BEEHIVE-SHAPED CHAMBERS WITH ELECTRIFIED AIR

This sounded so interesting that we brought our own magnetometer to the sites to try to confirm the readings of Cornet and Imbrogno. They were correct; the strength of the earth's magnetic field was always weaker right outside the entrances to the rock chambers.

There was no mystery about the source of these negative anomalies. The chambers were built in areas where magnetite is common. Magnetite is the magnetic form of iron – what the Chinese called lodestone, using it as compass needles. Incidentally, the rock chambers discussed in this chapter are located near a nineteenth century magnetite mine in Clarence Fahnestock State Park. A collapsed chamber lies on Magnetic Mine Road.

The shape of the interior of these stone structures is distinctive, like a beehive. Most of them are covered in a huge pile of dirt, and measure about fifteen feet across. Their walls of piled stones curve inward as they rise. Each succeeding layer of wall stones protrude slightly further into the airspace of the chamber until they reach a height of seven to eight feet on average. Here the walls stop, and table-sized slabs of stone lie flat over this abbreviated dome to form a ceiling. The archaeological term for this overlapping, inwardly protruding rock pattern is 'corbelled.' From inside, they look to us a bit like stone igloos with flat ceilings. From outside, they

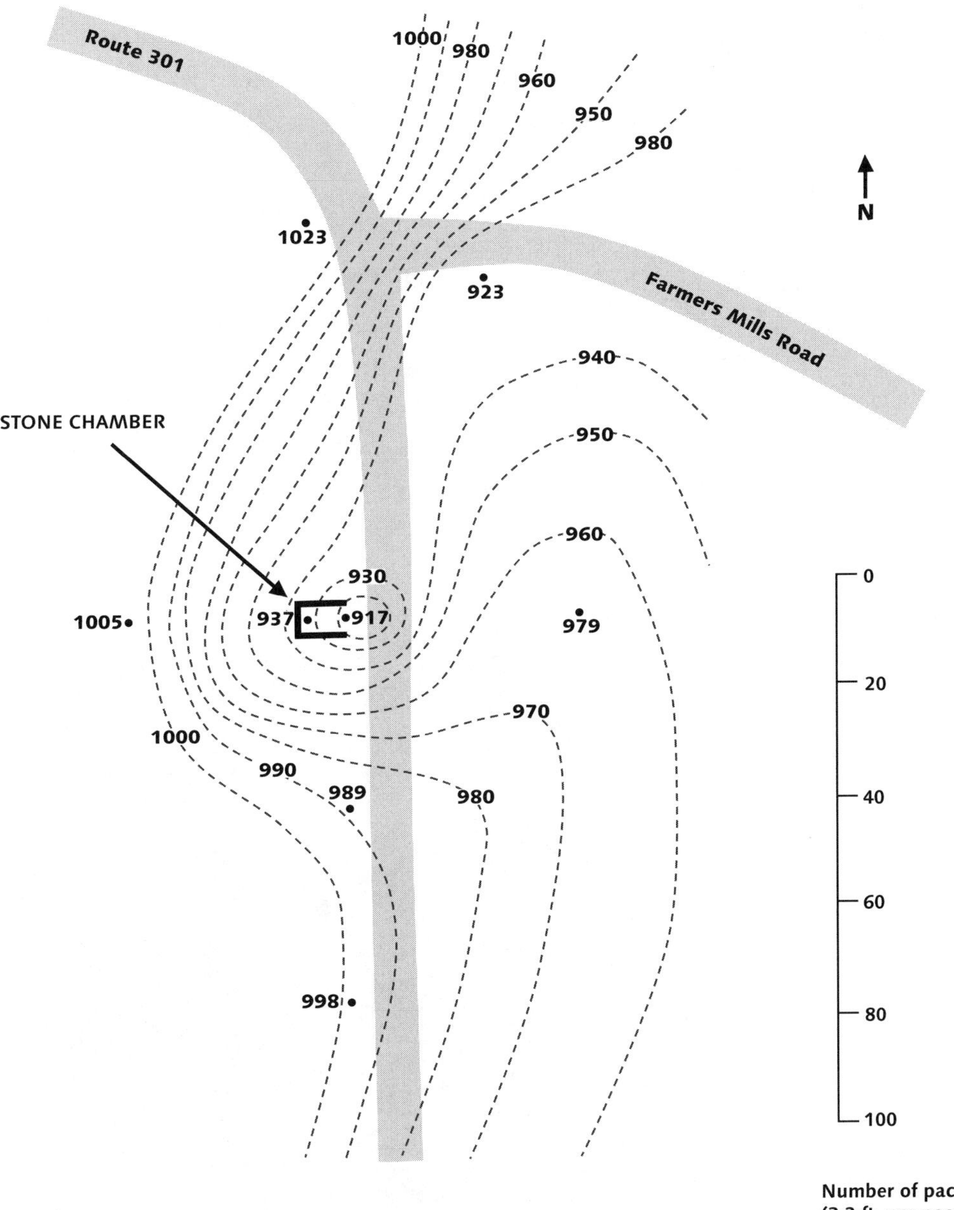

Figure 5. Geomagnetic contour map of a Native American rock chamber in Kent Cliffs, New York, showing a negative magnetic anomaly right at the open doorway – typical of such structures. This anomaly will electrify part of the air as it enters the chamber. (Courtesy of Dr. Bruce Cornet)

often just look like an inconspicuous pile of dirt having a rock wall with an opening on one side.

Some chambers remind us of an old-fashioned icehouse. Country estates used icehouses to store blocks of pond ice that had been cut in winter, to be used in the kitchen icebox. An icehouse had a conical roof, shaped ironically like an upside down ice cream cone, because this shape encourages lighter, hotter air to rise and heavier, colder air to sink, thereby keeping the ice blocks on the floor cool enough to last through the summer. We didn't think that the chambers were icehouses but wondered if perhaps gravity separation of something else was taking place here.

We found that the air inside these chambers was not at all normal – electrically speaking. The top few feet of air near the ceiling had a net negative electric charge compared to the more positively charged air near the floor (Figure 6). This is exactly the opposite of the normal fair-weather

Figure 6. Typical arrangement of electrified air molecules inside a Native American rock chamber. Heavier, positively charged ions sink to the floor, while lighter, negatively charged ions rise. This separation of charge sets up a measurable pulsing within the airspace of the chamber. (John Burke)

electric field of the Earth. Of the common, electrically charged air molecules (ions), the most prevalent are positively charged carbon dioxide, which is heavy and sinks, and much lighter, negatively charged oxygen molecules, which will rise above the carbon dioxide. Just like an icehouse separates air by temperature, so these beehive-shaped interiors separate, air ions by electric charge, in both cases due to differing weights.

Furthermore, the readings on our electrostatic voltmeter showed a persistent pulsing inside the chambers. The readings of electric charge of the air inside would suddenly jump, or fall. Sometimes the changes were small, at other times they were large – larger than usual in normal air, indoors or out. So how did this air get electrified in the first place?

Part of the answer is that the air is always electrified to some extent. But in magnetite country there is more to it than that. A basic principle of physics is called *induction*. It means that whenever anything moves through a zone of changing magnetic field strength, it acquires electric charge or current. Some of the air entering a rock chamber could be electrified just by moving across the changed magnetic field strength at the entrance. The faster it would move, the more powerful the effect would be. And we have found the air to be most electrified on those days that are windiest.

Was this electrical separation and its resultant pulsing the goal of the people who constructed these chambers? This question got our attention because the rock chambers began to look more and more like large stone versions of the apparatus that we were using to treat seeds in the laboratory. The essence of this technique is to put seed between two large, flat metal plates that are electrified with opposite charge: negative on top, positive on bottom, like the air in the chambers. Through a special technology, the charges of the plates create electromagnetic pulses in the air between them. As we saw in Chapter Two, seeds treated in this way are improved dramatically in a variety of ways, with the result that more food per acre is produced.

The similarities with the rock chambers were too great to ignore, so we went back and tried to find out more. On our second visit, we happened to be struck by beginner's luck while taking flash photos of one chamber at dusk. In Plate 17, the photo was taken from across the road, while Plate 18 shows a close-up of the area to the left of the doorway and a completely unexpected glowing column or arch of white light. If you look carefully at the arch, you can see an inner structure to it. The entire phenomenon is composed of separate, parallel rectangles of white light. Previously, we had found similar structures in other localities with high levels of natural electrical activity. We had also encountered them in other experiments on electrically charged air masses, called *plasmas*.[3] It was annoying – but not puzzling – when we realized that our electrostatic voltmeter had been 'fried.' Located in an outside pocket of the backpack shown in Plate 17, lying directly below the white arch in Plate 18, the unit had been subjected to a disabling electrical surge that left its needle stuck near the maximum reading of positive charge. It was six months before it worked again. From repeated experience, we would come to expect electrical equipment problems near the chambers; for example: batteries that would fail to work in a chamber, then work fine again when taken miles away, just to fail again in the next chamber.

ELECTROMAGNETIC SEED ENHANCEMENT

We were fascinated by the physical properties of these electromagnetic stone structures, so we put seeds inside several of them for differing periods of time, and then removed them. At Pinelandia Biophysical Laboratory in Michigan, Dr. W.C. Levengood germinated them in a standard germination chamber, using internationally accepted scientific protocols. We compared growth of seeds left in the rock chamber to those of control groups of seed that had been kept in our car at the chamber, and also back at the lab.

The seeds left in the chambers showed changes similar to what we had previously found in our lab, using the modern techniques. A larger percentage of them germinated. They germinated earlier and grew faster. Finally, they were far more uniform in growth than the controls. In these early tests, control lots of seed were also sometimes placed just outside the chambers, even at the spot where we had previously photographed the glowing arch. They did not show improvement. Something very special was happening inside the chambers.

Our experiment was repeated again and again over a two-year period. We consistently found that the chambers would improve seed and that the improvements were often statistically significant. In other words, some of the changes were so large that with a ninety-five percent probability, the unusual electromagnetic environment of the chamber had in fact done something profound to the biology of the seed. Sometimes the findings were at the ninety-nine percent probability level, meaning the odds were one hundred to one against this being the result of random chance.

During our first tests, we had no idea of how long to leave the seeds in a rock chamber to get good results. Some were left inside for two days, others for just an hour and a half. What happened made us realize that the timing was not a simple matter. Three of the first samples left in the chambers were improved to a level of statistical significance versus the control left in the car. However, a sample left for 50.5 hours was actually harmed to a level of statistical significance. This effect was similar to what we had found in our laboratory seed treatments. More of a good thing is not necessarily good. The 50.5-hour sample probably got an overdose of energy.

We would see this effect time and again, inside and outside the lab. Clearly, we were dealing with a natural balance here. Some of the effect was better than none, but it wasn't hard to get too much. On days when electrical activity around the chambers was particularly intense, we found that seeds left inside for thirty minutes outperformed those left in for sixty minutes, and both outperformed controls in the lab.

TESTING TRADITIONAL CROPS

We decided to try the types of seed grown locally in the days when the chambers were built. We located a type of traditional corn, which was grown up and down the East Coast from about 700 AD onward. We also wanted to try beans, because they were another major crop of Native American farmers. And finally we added winter wheat, because this is similar to emmer wheat, the staple of the civilizations that built almost identical chambers in Europe over thousands of years.

On October 14, 1995, we put winter wheat, some old type navy beans, and Tuscarora corn in two chambers simultaneously. All were improved! They not only germinated faster than the controls, but also grew faster under grow lights for a week, which was as far as they could be taken in the lab. For the beans, growth was actually tripled. Along with corn, beans were one of the staples of the North American diet. They were farmed along with squash long before corn was adopted from Mexico. When beans are combined with corn, they constitute a whole protein, which means the food can nourish a growing child or adult alike. Another experiment produced results that were particularly striking using low-vigor bean seed. The control seed that was kept in the lab had only fifteen percent of its seeds germinate compared to an average of eighty percent of the different groups of seed placed in two rock chambers. This advantage would have been of special value to the Indian farmers. The germination rates were so low that early European observers report that they planted five seeds in each hole to insure that one of them would germinate. This is substantial waste of what could have been used as food. In Europe, academics estimate that one quarter of the harvest had to be replanted as seed, again seriously denting the food supply.[4]

We began to notice, however, that determining the optimal length of the treatment was not a simple matter. For the winter wheat, 105 minutes in the chamber improved the seed more than a thirty-minute stay. For the beans, two hours was better than one hour. For Tuscarora corn, on the other hand, thirty minutes was better than three hours. We began to equate this to cooking meat in an oven. Each type of meat does best at a different temperature. Then even the ideal temperature must be maintained for differing lengths of time, depending on whether you are cooking chicken or roast beef. If you have an oven without an accurate thermostat, you have to alter the cooking time. But by how much? If you also lack a meat thermometer, a certain amount of guesswork becomes involved, an intuitive 'feel' for when the roast is ready. This feeling is rooted in experience. A similar intuition was probably involved in using the rock chambers to enhance seed.

Of the two rock chambers used in our early tests, we found that either one could outperform the other on a given day. In fact, one chamber might produce no changes in the seed, while the other would improve them dramatically. This outcome made perfect sense because our readings of electrical separation in the air inside the chamber also varied from day to day. In fact, we began to be able to predict what constituted a 'good' day. If there was substantial separation of charge in the air inside, and if the electrical readings varied in frequent pulses of significant size, we could generally count on a thirty-minute exposure improving the corn. When we found no separation of charge in the air inside, typically in the winter, we would see no effect on the seed at all. When there was no significant pulsing, we got no effect. We cannot emphasize enough that it was only on rare occasions that the effect on the seeds was negative.

Before long, we were shown a startling infrared photo taken inside a rock chamber in Ninham Mountain State Park, New York. A glowing mass, shown in Plate 19, was suspended in midair. It was invisible to the naked eye but showed up on the infrared photo. While it is still unknown to science what could account for this light, similar reports do exist in the scientific literature.[5] Our magnetometer survey of the chamber showed that it was located on the usual negative magnetic anomaly. In addition, it lay on a boundary where a zone of homogeneous magnetic readings downhill from the chamber met a zone of highly varied readings uphill, a pattern we would see again and again. The roof slabs directly over the glowing ball were some of the most magnetic rocks we had ever encountered anywhere. Some careful magnetic engineering seemed to have taken place here hundreds of years ago.

In Chapter Three, we learned that most people can be quite sensitive to minute changes in these natural forces. An average person can detect the magnetic anomalies we found under the right conditions. And some people are far more sensitive than others. These chambers were built when the area was inhabited by the Wappinger and Mohegan tribes. Perhaps the Native American farmers, or at least their shamans, discovered the connection between energy and seed.

The following year, we received Iroquois Blue Flint corn seed from Lawrence Davis-Hollander of the Native Seed Conservancy in Great Barrington, Massachusetts. Lawrence got his Ph.D. at Harvard in an unusual field. Ethnobotany is the science of studying plants grown by ancient peoples, and keeping those seed stocks alive. We are grateful to his mentor at Harvard, who convinced Lawrence that if someone did not do this, these old plant types would be lost forever. He launched Lawrence on a pioneering road that has shown just how important ancient knowledge can be. Lawrence now spends his time doing things like hunting up farmers on New York's Mohawk reservation, a few of whom are farming particular varieties of corn and beans that are no longer in commercial use. Then he will hunt up someone else and turn on his considerable charm to convince them to grow these varieties and give him some of the next generation of seed.

On a single day, we put three samples each of Blue Flint and Tuscarora corn inside three rock chambers. At all three sites, control seed lots were placed 100 feet outside the chamber, while other controls were kept in the car and in the lab. Then the Blue Flint seed was planted and the controls performed consistently similar to one another. However, the nine lots that had been left in the three chambers had their yields raised in proportion to how long they had been inside.

Some of the Blue Flint corn was placed inside the rock chamber where the energy mass in Plate 19 was photographed. Electrostatic voltmeter measurements of the air inside the chamber on the day the seeds were exposed had shown one spot to be much more negatively charged than any other. This was the spot where the glowing ball had appeared in the infrared photo. We placed samples of the Blue Flint seed on the dirt floor there for seventy-three minutes. When compared to the five control samples, the sample left longest on that spot in the chamber had eighty-four percent more seeds germinate and grow to maturity.

The Blue Flint corn was grown to maturity in the field by Lawrence and his assistants, using Native American farming methods. While I was on a follow-up visit shortly after planting, Lawrence's assistant rushed in, flushed with excitement. He blurted out that one row of the Iroquois Blue (the row left longest in the chamber) was up, and they had never seen Iroqouis Blue emerge like this before. We hurried out to the field where it turned out that ninety percent of this row was up, uniform in height, and looking vigorous. Normally, only fifty percent of the seed germinates. In addition to the obvious advantage of a higher percentage of germination, speed of emergence is of great value to every farmer, because in the early stage of a plant's life, it is most fragile, and the sooner it gets through this stage the better. Equally important was the fact that the chamber seeds grew far more uniformly. As a result, most of them would mature at the same time. This is a property of extreme importance to farmers, because if a plant reaches sexual maturity too far ahead of or behind its neighbors, it will have no way to pollinate or be pollinated. In that case, it may grow to full height but never produce an ear. Even with the best modern corn seed, this still happens to a percentage of plants on today's farms. A few centuries ago, the problem was much worse.

Both the Blue Flint and Tuscarora corn seeds left in the chambers did in fact have dramatically fewer free radicals than the control seeds

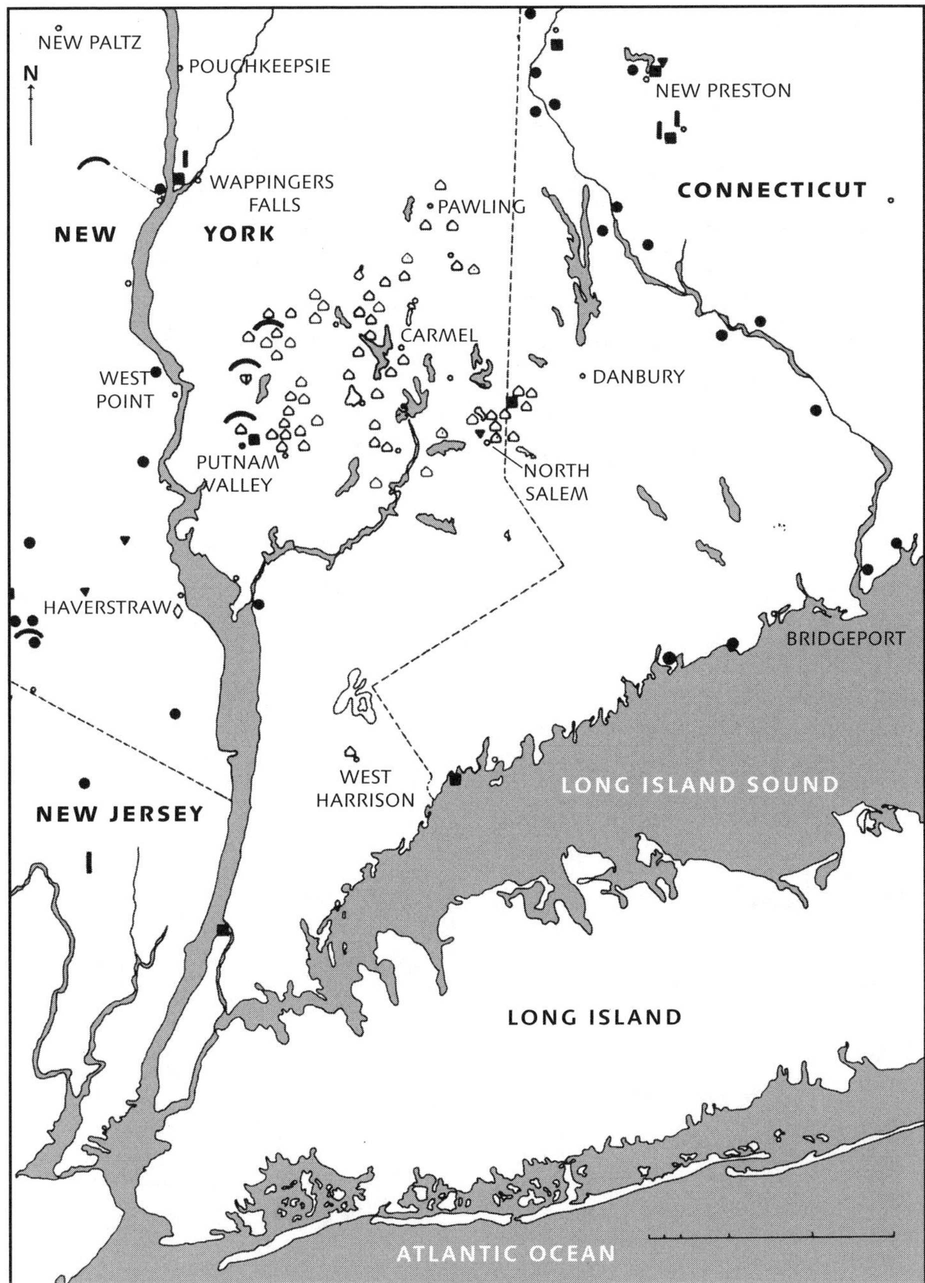

Figure 7. Map of pre-Columbian rock chambers in southern New England. Most of the largest chambers in the United States are found in this region, including the complex at Mystery Hill, 'America's Stonehenge,' in North Salem, New Hampshire. (Illustration from Salvatore Trento. *The Search for Lost America*. Contemporary Books, Chicago, 1978, by permission)

when tested by bio-electrode analysis.[6, 7] We expected improvements in the harvest, but the differences exceeded our expectations. As shown in Plate 20, forty seeds left in the chamber produced three times the amount of corn as forty seeds left a hundred feet outside the chamber. Forty seeds left for only thirty minutes in the chamber with the glowing plasma arch (Plate 18), produced more than twice the amount of corn as forty seeds left in the lab. Of the nine chamber entries, all fell within this range. Apparently, if a Native American farmer placed his seed inside a rock chamber and left it there for an hour or two, he could have doubled or tripled his harvest.

NATIVE AMERICAN ENGINEERS

Perhaps this is the reason hundreds, if not thousands, of these rock chambers are spread all over the northeastern United States[8] (Figure 7). Appendix 1 lists locations of numerous locations open to the public today. More than a hundred cluster between the New York towns of Brewster, Mahopac, and Kent Cliffs. Woodstock, Vermont has striking examples, along with southwestern Connecticut. The Gungywamp complex, in southeastern Connecticut, is examined in Chapter Three.

Probably, the most famous group of rock chambers is at Mystery Hill, often called 'America's Stonehenge,' in North Salem, New Hampshire. Here, carbon-14 dating has shown that the rock chambers originated long before Columbus. In a few cases, organic remains have been discovered inside the chambers, the oldest dating back to about 200 AD.[9] We brought our magnetometer to Mystery Hill to find that the chambers here also have negative magnetic anomalies at their doors.

We are disappointed indeed that these chambers have been largely ignored by American archaeologists. In fact, academics seem to have gone out of their way not to study them. The excuses have been varied, including a dismissal of the chambers as root cellars or storage sheds of colonial farmers. This interpretation is inconsistent with the following facts: The chambers were reported by the first white settlers of some areas;[10, 11] root cellars are not built at ground level, where everything freezes solid in the winter; a colonial farmer would most likely not invest what one engineer who owns a rock chamber has carefully estimated to be six months of labor with a horse to drag tons of stones downhill, and use the help of gravity to slide them into place, when that same farmer could have built a log storage shed with a small fraction of the time and effort.

We understand that the issue has been clouded by a few chambers that were later modified. For example, in Ninham Mountain State Park, a farmer added cement and two-by-fours to the opening of a nearby chamber to put in a door and use the chamber as a storage shed. But such chambers are the exception. In fact, many of the most impressive chambers lie in marshes like Gungywamp or on mountains in Woodstock, Vermont, where no one ever farmed and no colonial settler lived.

Some organizations that study the chambers, note that they are virtually identical to dolmens found all over Europe. Their interpretations of the chambers' origins seek to make a connection between the chambers and dolmens. These theories have ranged from Druids and Vikings to Irish monk explorers, Phoenician and Portuguese sailors. Old World residents did erect virtually identical structures. Over a 6,000-year period, thousands were erected from Ireland and Britain, through France, Portugal and Spain, and eastwards as far as India[12] (Figure 8). Dolmens are a phenomenon linked to many cultures. Given what we now know about the capabilities of the chambers, we consider it likely that they were utilitarian structures, serving an eminently practical purpose – rather than being solely ritual or cultural -- and that this was the cause of their almost worldwide distribution. In Chapters Eight and Nine, we examine the evidence to back up this theory. Whether their invention arose separately or was transmitted by early explorers, we propose that it was their practical benefits that caused so many to be built. The similarities in design, therefore, could either be from knowledge passed from one people to another or from the engineering necessary to achieve the desired effect.

STRANGE SENSATIONS AT BALANCED ROCK

Balanced Rock is a marked historic site located on Route 116 in North Salem, New York (Plate 21 and Figure 9). This room-sized, over sixty-ton boulder is perched precariously atop several small, white slabs of granular quartz that rise vertically from the ground and give all the appearance of having originally enclosed a small space beneath the giant rock. At what looks like the door of this chamber, our magnetometer revealed an unusually strong magnetic anomaly. This spot is the intersection of two different areas of homogeneous magnetic fields. On one side of the rock, the land for some distance around has a vertical field strength of 54,200 gammas. On the other side, the land all reads 54,400 gammas. But in the

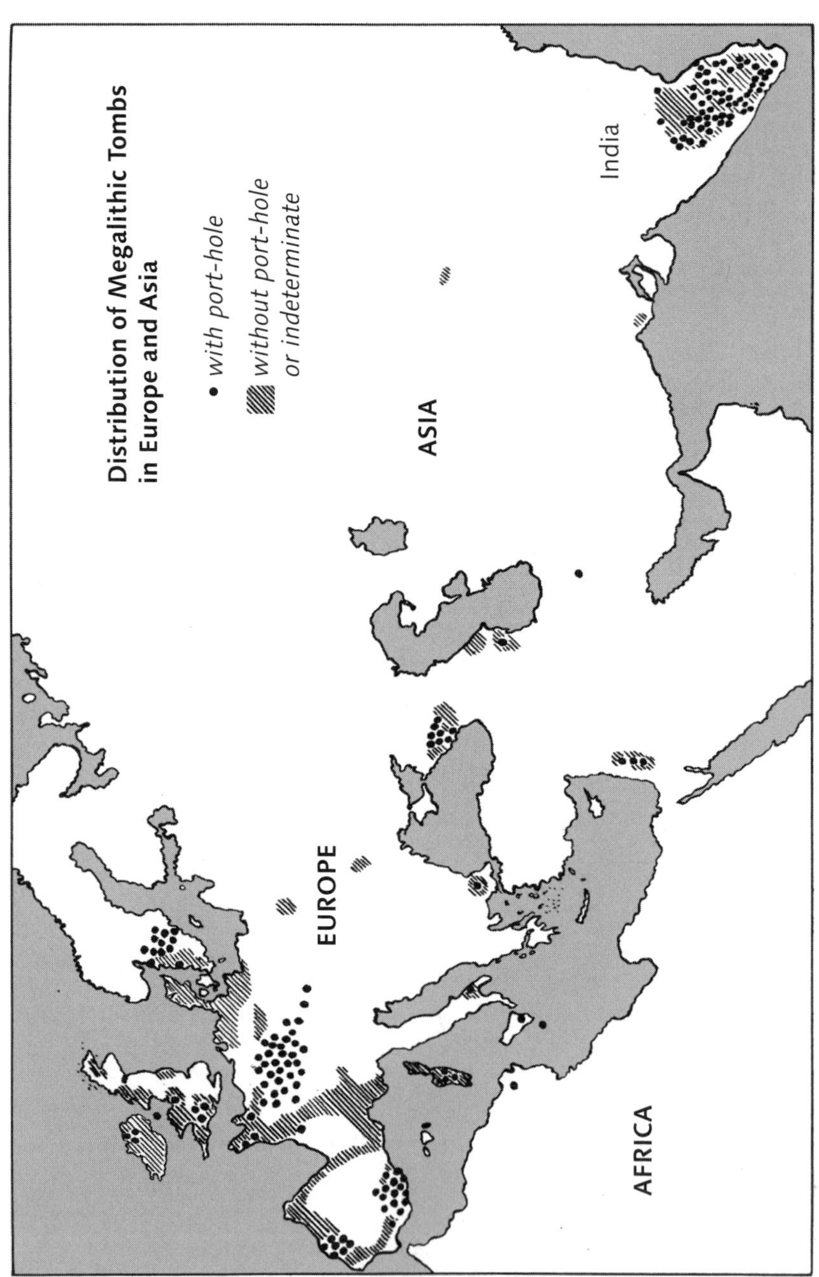

Figure 8. Rock chambers, or dolmens of identical design, were built throughout the world. They always first appeared after a community had been farming for several centuries, thereby exhausting the soil. Perhaps the stone structures were designed to enhance seed performance and compensate for deteriorating soil in times before crop rotation and fertilizer were known. (Illustration from Joseph Campbell. *The Mythic Image*. Princeton University Press, 1981, Map 6, by permission)

middle, right where the rock is perched, the vertical field strength plunges to 53,800 gammas. This anomaly is not located in the boulder itself but in the ground beneath.

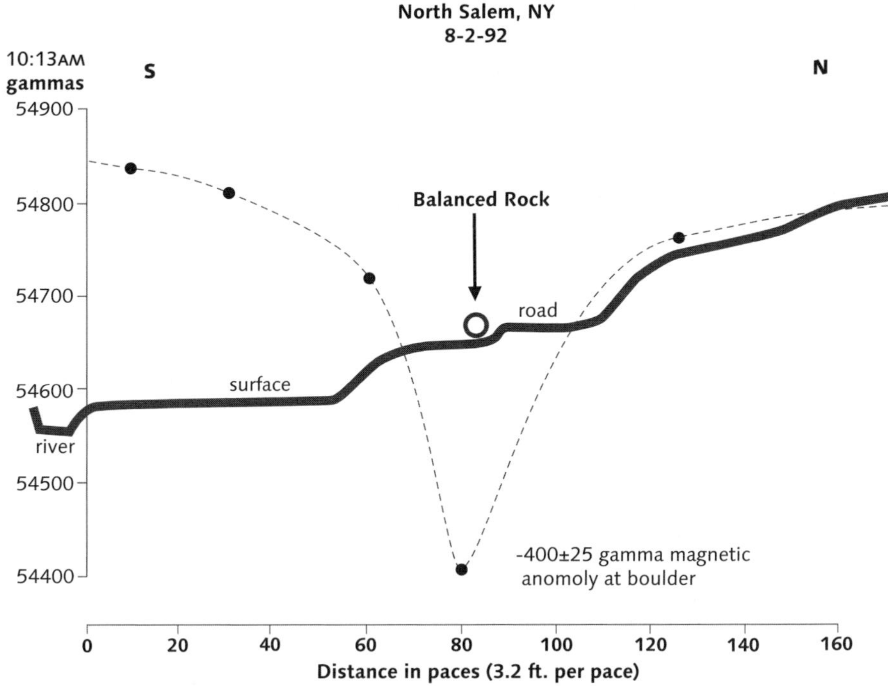

Figure 9. Balanced Rock (shown in Plate 21) is a 90-ton boulder, perched precariously atop a few upright quartz slabs, directly above a 400-gamma magnetic anomaly. From the spot in the ground directly below the huge boulder, geomagnetic readings rise sharply in all directions. (Courtesy of Dr. Bruce Cornet)

The Earth's magnetic field is quite uniform everywhere, in percentage terms. A 400-gamma magnetic difference is huge for natural objects. Golf-ball sized globes of orange light have frequently been photographed at Balanced Rock. Numerous people, who sat up the night at the rock, have reported strange sensations and perceptions. It is not surprising to us that Native Americans were able to recognize this place as special. The historical marker here labels the rock as a glacial erratic. It seems improbable though that it would have accidentally settled on several connected upright quartz slabs on the exact spot of a striking magnetic anomaly. American archaeologists generally do not pursue the theory that pre-Columbian technology could have moved such a boulder. But Balanced Rock is no larger than many of the dolmen capstones in Europe, some of which weigh up to a hundred tons.

FERTILITY CAVES

Wondering how the Native American builders might have conceived of these pulsing 'fertility generators,' we were intrigued to hear of the long history of caves in the Meso-American cultures that preceded the North American chambers. The Olmec and Mayan cultures (Chapters One and Four) were the first to start large-scale maize and bean farming, making it possible to give up a hunter-gatherer existence. Their knowledge spread north to eventually reach tribes of the present-day United States.

In the Tuxtla Mountains of Mexico, the Olmec would often store seed in caves. Perhaps in this way, they first accidentally discovered the fertilizing powers of the electromagnetic energy of Mother Earth. Like their Olmec predecessors, the Maya regarded caves at the peak and foot of sacred mountains as the origins of fertility, and an opening framed by human lower jaw bones, or mandibles, was a symbol of this fertility cave.[13] (In Europe, lower jaw bones were buried in great numbers at megaliths and henges, as we shall see in Chapters Eight and Nine.)

Specially marked 'fertility' pottery full of maize was often presented in certain Mayan caves as offerings to the jaguar fertility god.[14] Such offerings were left there a while, then removed.[15] If these chosen caves were anything like the New England rock chambers, this would be a very effective procedure when the energies were right. The result would be seed with an ability to produce several times more food per acre. Certainly the Maya believed that enhanced fertility was obtained by these rituals. In order to commemorate these powers they built artificial 'maize mountains' – what we call pyramids – near their sacred caves.[16]

As we contemplate the prevalence of beehive-shaped rock chambers from North America to Ireland to India to Peru, we can't help but be reminded that the Mayan glyph for fertility is a cave with a beehive inside.[17]

When we first stood in Lawrence Davis-Hollander's field and gazed down at our newly germinated seedlings, we were reminded of the Mayan myth for the origin of corn. It is described in the sacred book of the Quiche Maya, the *Popol Vuh*, the sole surviving script of this people, written shortly after their capital was destroyed by Spanish conquistadors in 1524.[18] According to this book, maize lay undiscovered in a mountain cave underneath a large rock until a bolt of lightning penetrated the cave and split the rock apart, revealing the seed of the crop that was to become the staple diet of tribes throughout the Americas.[19]

CHAPTER 7

Mystery Mounds of North America

"The chief role of American Indian religion was to help the crops grow better."

— Ake Hultkrantz[1]

So here I am, in sub-zero wind chill, standing atop a huge earthen mound at Cahokia, Illinois. To properly make magnetometer readings, I repeatedly have to stand completely still for several minutes at a stretch. My body begins to rebel, I shiver from the cold, and two or three toes are frozen stiff. What in the world am I doing here?

To answer that question, we must go back to the times when corn farming spread northwards from Meso-American cultures, spreading certain other aspects of these cultures with it. Throughout the middle portion of the United States, thousands of Native American earthen mounds still stand, hugging their histories close to their hearts. I was trying to tease out these histories, and frostbite nearly cut my attempt short.

THE MOUND BUILDERS

About 2,400 years ago, in present-day Ohio, a group of hunters and primitive farmers that we today call the Adena began to heap up earth into mounds. The Adena were not a tribe, but part of a culture that embraced many tribes in the area. Several centuries later, they were succeeded by the people of the Hopewell Culture, who were accomplished farmers. Their influence reached down the full length of the Ohio and Mississippi Rivers.

The center of the early Adena and Hopewell mound building was around the modern town of Chillicothe, Ohio. United States Geological Survey magnetic maps show that the town and its surroundings lie between two large regions of powerful magnetic fields to the north and south. Running east-west between these areas is Paint Creek, along which countless mounds and earthworks were erected. Almost none were placed further north or south, outside this magnetic transition zone. In Hopewell Culture National Historical Park at Chillicothe, a few dozen mounds survive on the flood plain of Paint Creek. Every mound over six or eight feet high was placed on a single magnetic anomaly strong enough to be detected by our magnetometer. The mounds are round, with one exception. Where there had been a line of positive magnetic anomalies in the ground, the Hopewell builders erected a linear mound, whose ridge runs across all the anomalies. The largest group of these mounds is surrounded by a henge-like ditch.

Someone had sought out spots with powerful magnetic anomalies and decided to build mounds there. Many Hopewell and Adena sites were anything but modest piles of dirt. Some stood several stories high and contained hundreds of thousand of cubic feet of earth – all laboriously heaped up by hand. Archaeological excavations have shown that these structures were built by people scraping earth into baskets that would be carried on their back to the site, up the existing mound, on top of which it was dumped – repeating the process endlessly.[2]

Mound building in North America was brought to its zenith circa 800-1400 AD by the Mississippian Culture, named for the huge river that was their center of activity. During a period of population density that was unparalleled in pre-Columbian North America, the Mississippians built the largest earthen mounds the country has ever seen. At their great center at Cahokia, Illinois, they raised a giant, flat-topped pyramid, rivaling in volume the Great Pyramid of Giza in Egypt, as well as a few dozen smaller mounds (Plate 22). For several hundred years, Cahokia was the most spectacular settlement north of Mexico. At its peak, about 1300 AD, it was home to thirty thousand people, exceeding the population of London at the time.[3] Not until Philadelphia around 1800 AD, did an American city grow larger.

At least 383 other villages of this culture bordered the Mississippi River between the Ohio and Red Rivers, and thousands of others spread up and down its tributary streams, as far afield as the Missouri River in South Dakota. Though a few mounds contained some skeletons, most had nothing to do with burial. Yet these mind-boggling, laborious creations were built in the hundreds throughout this region by different tribes using different languages (Figure 10). Appendix 1 lists locations of numerous examples open to the public today.

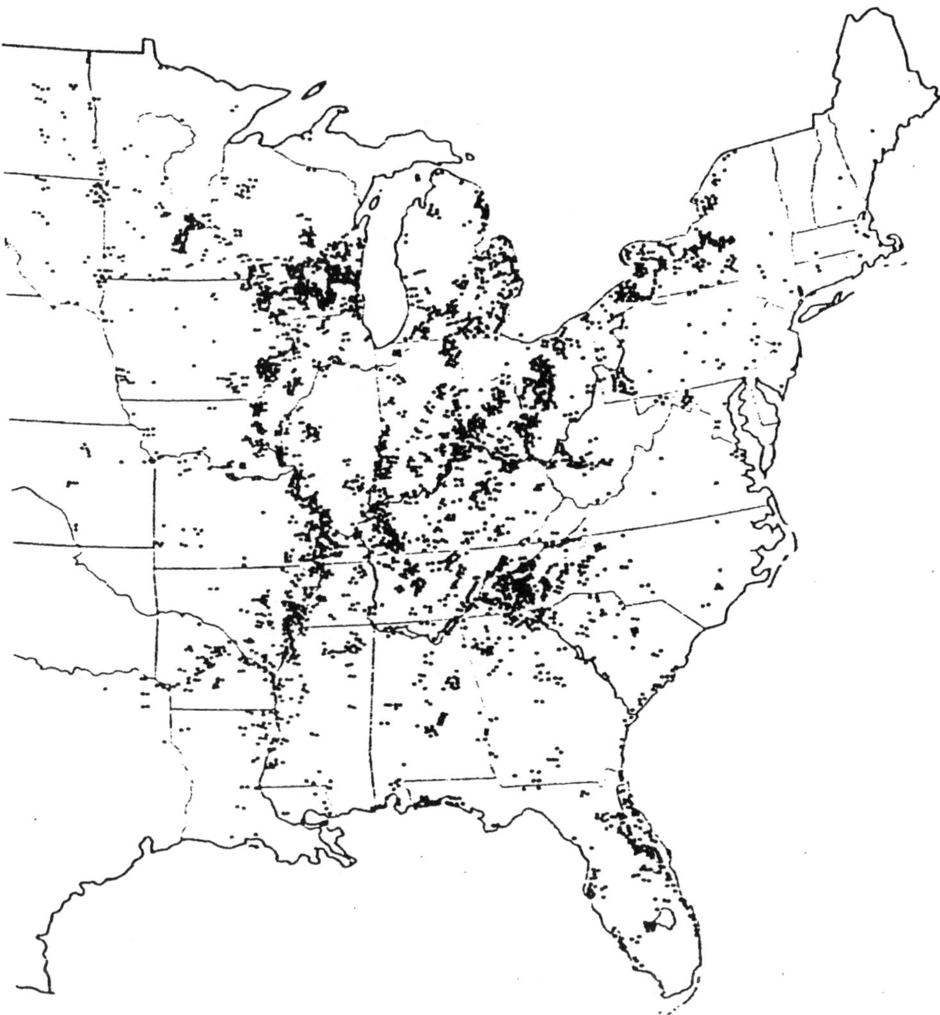

Figure 10. Mounds in uncounted thousands cover much of the United States. While many mounds contained bodies, others were clearly intended for other, unknown purposes. Ranging in height from a few feet to 100 feet, many are still preserved in state and national parks. To locate the ones nearest you, see Appendix 1.

Incredibly, in most cases the builders did not even use the dirt closest to the mound. They usually went some distance for special clays that were then applied in layers, one type at a time. When you consider that these local populations usually numbered no more than a few thousand people, what would compel them to dedicate hundreds of thousands of man-hours of backbreaking work to haul clays from a distance, only to dump them out and pile them up? What made these mounds worth such staggering effort?

Like the Adena, the Mississippian Culture was not one tribe but a way of life that included many tribes. One characteristic of the Mississippian Culture was the intensive farming methods imported from Meso-America; for example, growing two crops of 120-day-corn per year in the same field.[4] Even today, no farmer double-crops like this unless there is a serious need to get more food from the land. For one thing, there are substantial risks in double-cropping. If one of the plantings experiences unusually cool weather, then those plants will take longer to reach harvest maturity, posing the risk that the reduced number of days now left for the second planting to mature may be insufficient. Even today, this is not an uncommon experience for double-croppers, and the result is that the second planting does not mature before winter and the entire crop is lost. Were the mounds partly built for this very reason? Were they constructed in a way that would improve the seed in the same manner as is accomplished with the modern electron showers that we discussed in Chapter Two? If this were so, then the Mississippian crops would have matured sooner, making it more likely that both 120-day-corn crops could reach full maturity. As we know from the previous chapter and our experiments with these energies in the New England rock chambers, the additional value would be that both crops would produce substantially higher yields.

MOUNDS, ELECTRICITY, AND FERTILITY

Careful surveys with our magnetometer at seventeen mounds showed a persistent connection to positive magnetic anomalies; and in every case, the readings were dramatically higher at the top of a mound. The sides might also give higher readings than the surrounding flat ground, which showed no magnetic anomalies, but the highest readings were invariably at the highest point. And the larger the mound, the bigger the difference.

Archaeological excavations had shown that the mound itself was not magnetic, so it had to be the ground under the mound. The creators had decided to build the mound directly on a magnetic anomaly.

Our surveys of the electric charge in the air above the larger mounds produced readings that were some of the highest we had ever encountered in nature. Both these and the magnetic readings varied widely between nighttime and midday. Given all this data, it seems clear that the building of these mounds probably had something to do with electromagnetic energy.

We had brought a set of ground electrodes to detect telluric currents in the earth. Such ground currents are normally quite steady, but at the mounds, they were anything but. Here, they sometimes pulsed in rough unison with the magnetic and airborne electrostatic changes in a way that was consistent with known physics, just not normally seen.

The river connection is relevant. The vast majority of major Native American mounds are located on high ground bordering a river. Granted, the rivers were the highways. But English henges (as we shall see in Chapter Nine) were located in the same type of locations. Furthermore, both the American and the English builders ran 'avenues' lined by ditches from water's edge to the mound, for reasons that we shall revisit.

The famous Serpent Mound in Ohio is a quarter mile long embankment built by Native Americans in the shape of a writhing snake (Figure 11). Long credited to the Adena, circa 500 BC, recent carbon-14 dating has proven that it was actually engineered by the later master mound builders, the Mississippians, around 1070 AD.[5] This combination got our attention because in the Mississippian Culture, the snake symbolized fertility. In fact, in their mythology, the serpent was considered the source of all plant life.[6] A frequently used symbol of fertility was that of a serpent rising out of the ground in the form of a plant.[7]

On our first visit to Serpent Mound, we encountered dramatic readings. Heavy gusts of wind, charged to one thousand volts per inch, were blowing in from the northeast and the direction of other native mounds. This might sound fatal, but the current is so low it has little impact on people. We thought, though, that it might have an impact on seeds.

On the side of Serpent Mound, we placed seeds in porous cotton gauze bags – but were disappointed. While the seeds showed the same type of

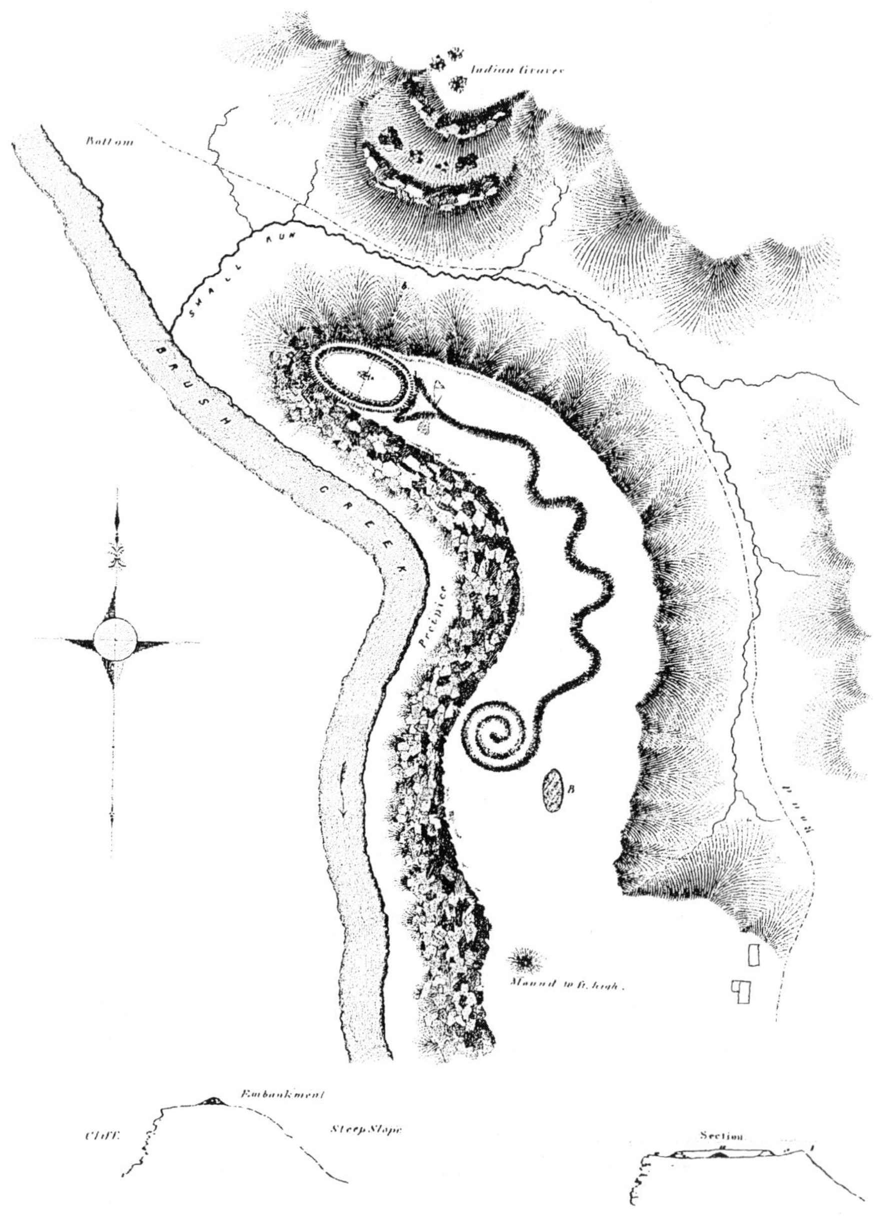

Figure 11. The quarter mile long Serpent Mound in Ohio was built inside a highly unusual geological structure known as a 'crypto-explosion crater.' During an approaching thunderstorm, electromagnetic readings here were the most dramatic ever recorded by the authors. Seeds placed on the mound at the time later exhibited dramatically enhanced growth. (Illustration from E.G. Squier & E.H. Davis. *Ancient Monuments of the Mississippi Valley: Comprising the Results of Extensive Original Surveys and Explorations*. Smithsonian Institution, Washington DC, 1848)

improvements that we had associated with New England rock chambers, the differences were small. The seeds left on the mound grew only eight percent faster than the control seeds. Were we missing something?

In the mythologies of mound-builder cultures throughout Meso-America and the U.S., man-made rock chambers and natural caves were linked with mounds, fertility, corn, seeds, water, and one other ingredient – lightning.[8] We had worked with all of these factors but the last. One day, at the sixty-four-foot-tall Grave Creek Mound in West Virginia, after the passing of a thunderstorm, the telluric currents reached an astonishing three volts/km, the highest we have ever recorded anywhere. In the flat ground surrounding the mound, the currents were not even near that level.

Then, in August 1996, we brought seeds of corn and beans to Serpent Mound, on a day when a thunderstorm struck. There is a certain poetry to this when you consider that in Mayan literature like the classic *Popol Vuh* – and also among Mayan ethnic groups today – the serpent symbolizes lightning,[9] while in the Mississippian mound-building culture, it represented fertility. At three different locations on the writhing flanks of the mound, we placed bags of three different types of seed for differing lengths of time. The thunderstorm moved in, and ultimately lightning got to within one or two miles of the mound. During its two-hour approach, something amazing was happening.

From the time we began taking readings on our magnetometer, at 2:06 PM, the readings plunged. Starting at a normal fair-weather level, the magnetometer readings dropped several times further than we had ever before measured anywhere. This drama continued as the lightning got closer. In near disbelief, we installed a new battery, but that didn't make any difference. Something radical was happening to the magnetic field. Ultimately, we had to cease measuring and retrieve the last of the seed, when downpours began at 3:57 PM.

There is no doubt that these dramatic magnetic fluctuations were produced by powerful electric currents, streaming through the bluff under Serpent Mound. Thunderstorms are known to induce such currents in the earth; without them, lightning would never strike. These telluric currents alter the local magnetic field by producing a magnetic field of their own. If the new field is opposed to the normal one, it will weaken it and thereby lower magnetometer readings. It was obvious that beneath Serpent Mound, the telluric currents were coursing through the ninety-foot bluff that sticks out into the valley like a peninsula. It is composed of solid dolomite, a particularly electrically conductive form of limestone with a high content of magnesium.

We knew that the terrain was unusual. Flanked by two streams, Serpent Mound sits within a strange geological structure called a *crypto-explosion crater.* Millions of years ago, a five-mile-wide bowl was carved out of the earth here by a massive explosion. The entire center of the crater is one large negative magnetic anomaly. Near the center sits the dolomite bluff, which was artificially flattened by the mound builders, who left only a thin covering of soil atop the dolomite. Then the serpent was laid down with alternating layers of clay, carried up ninety feet from the stream beds below. On one side of this bluff is a small cave, located just below the serpent's head.[10]

With the cave, the mound, the serpent, the seed, and lightning, we now had all of the elements of mound-builder fertility imagery present in a real, physical form. What happened to the seeds placed on the mound

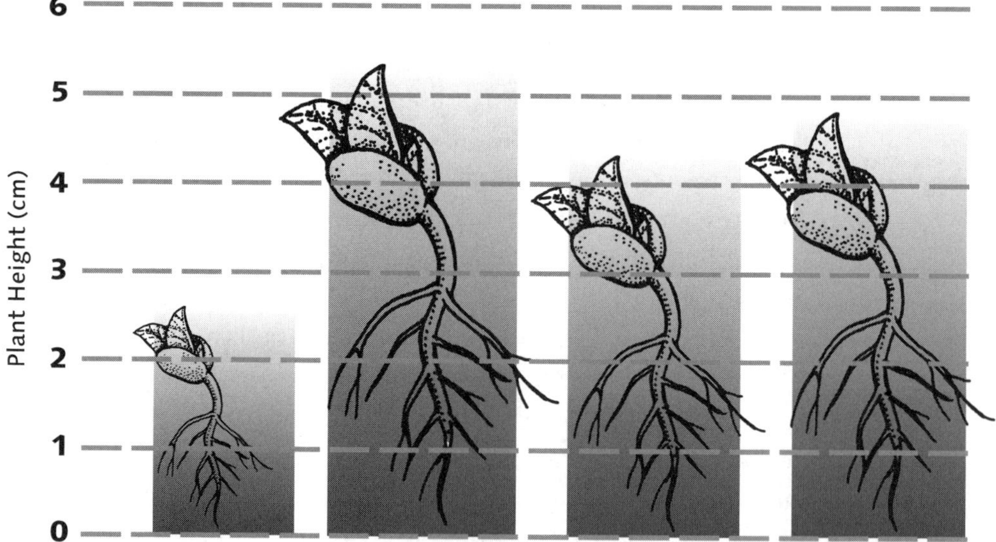

Figure 12. Seedling growth in navy beans placed on Ohio's Serpent Mound during an approaching thunderstorm showed much greater average height and uniformity compared to an identical seed sample kept in the lab as a control, see the bar on the far left. (Courtesy of Pinelandia Biophysical Laboratory, Michigan)

left no doubt at all that the 'fertility' of Maya myth was also present in physical form. As shown in Figure 12, the seeds placed on the mound grew dramatically faster and more uniformly than the control lots kept in our car near the mound or those left behind in the lab. The differences were statistically significant.

The improved growth closely correlated with the changing telluric ground currents and their magnetic fields. The least growth improvement was seen in the seeds placed on Serpent Mound before the electromagnetic changes got very large. The greatest growth increases were attained in the seeds placed on the side of the Serpent, at the height of the electrical and magnetic disturbance, growth improvements similar to our results atop the Lost World Pyramid (Chapter Four).

THE TRADE DEFICIT

Dr. Brad Lepper is an Ohio boy who rose to the top of his field, pursuing the question that had fascinated him since childhood: How did the Hopewell mound builders rise from a scattering of obscure tribes, circa 800 AD, to create a high civilization east of the Mississippi? It wasn't just the mounds. They also dug ditches that bear striking resemblance to the henges of England. In Ohio, these ditches were often created in geometric patterns, enclosing dozens of acres and, as we shall see, concentrating electric ground currents. How did the Hopewell do this, and how did they become wealthy beyond any others who came before them?

Today, Dr. Lepper is the State Archaeologist of Ohio. The work that put him in this position includes papers published in peer-reviewed journals, citing a fundamental puzzle. The Hopewell wealth shows up in grave goods and excavations in the form of luxury items, such as mica from the Rocky Mountains, copper from northern Michigan, seashells from the Gulf of Mexico, and more.[11] But try as he might, Dr. Lepper could not find what the Hopewell themselves exported to pay for these items. Well, they did make fine flint hoes that have turned up in archaeological sites up to a hundred miles away, but not in sufficient quantities to have counterbalanced the expensive imported items. "There is a missing factor here," he told us.[12] This reminded us of a similar statement made by two Olmec archaeologists. For the Hopewell, as for the Olmec, the only thing they seemed to possess, that others did not, were mounds.

'GIZA' IN DIRT

This 'missing factor' would reach its zenith circa 1000 AD in the earlier-mentioned trans-tribal culture called the Mississippian, whose center was at Cahokia, Illinois, just across the Mississippi from present-day St. Louis. I went there with our instruments, determined to get some answers. However, I arrived in winter – and soon came to wish I hadn't. Many of the electromagnetic forces discussed in this book are strongest in warmer weather. The rock chambers, described in Chapter Six, often show no charge separation and no pulsing in temperatures below freezing. The abilities of other factors, involving interactions of chalk and limestone with water are also temperature dependent. However, magnetic anomalies are usually permanent and can be measured any time of the year.

In the Hopewell Culture, the bigger the magnetic anomaly, the larger they would build the mound above it. So, trudging through several inches of snow in Cahokia State Park, I definitely had high expectations of what would pop up on our magnetometer. Facing a bitter winter wind, I struggled to the top of the biggest earthen mound in the world, the so-called Monk's Mound – at 22,000,000 cubic feet having a volume almost equal to that of the Khufu pyramid in Egypt.

This massive mound rises one hundred feet above the Illinois plain. Standing up here in the snow, exploring a new frontier of knowledge, I gazed across the river to St. Louis and its magnificent arch – a memorial to the pioneers who went west from here to explore and inhabit a different frontier.

THE MAGNETIC PYRAMID GRID

So what did I find atop this huge mound? Nothing. All across the top, the magnetometer readings were the same. To add insult to injury, I encountered an unexpected difficulty on the fourth reading. The LCD readout had frozen and refused to show the numbers. To properly read a magnetometer, a person must stand completely still for several minutes. In sub-zero wind chill, the body begins to rebel, and I was shivering from the cold. Over the next two-and-a-half days, I developed a routine: take a reading; stuff the magnetometer inside my down jacket against my chest to keep the readout from freezing; walk a hundred yards as briskly as possible; stop; unzip down jacket and remove magnetometer; try to stand still and not shiver; repeat.

Hundreds of repetitions later, by the second day, my efforts paid off. A geographic pattern emerged from the readings, one that would bear out and ultimately answer some of the long-asked questions about this mysterious site. Here, dozens of mounds – totaling sixty million cubic feet in volume – were laid down on a 'magnetic grid.' The one square mile that contains the major structures of Cahokia lies at a transition zone between an extensive region of lower magnetism (54,000 gammas) to the north and west, and a large area of higher magnetism (54,400 gammas) to the south and east. Though this difference may sound small as a percentage of the geomagnetic field strength, it is actually a large difference when you are analyzing the earth's field. This difference makes the entire Cahokia site a conductivity discontinuity – in other words, it is the meeting place of two regions that have differing abilities to conduct electric ground current. As we have seen, such spots tend to be especially powerful for electromagnetic energies.

In Chapter Two, we discussed the principle of induction: that changing magnetic field strength will create electric current in anything present that will conduct electricity. The same principle is true for the ground. At dawn, the changing strength of the geomagnetic field generates telluric currents that race through the ground, near the surface. Again, nature is rarely simple. These telluric currents do what any electric current does: They generate their own magnetic fields that will reinforce or weaken the geomagnetic field at that spot, depending on which way they are polarized.

As we have seen in Chapter Two, these natural geological sites, called conductivity discontinuities, can magnify the daily geomagnetic fluctuations several fold. And, since their fluctuations generate electric currents, the normal daily telluric currents at such a spot can be magnified by several fold.[13] Finally, a negatively charged telluric current in the ground will attract positively charged particles and fields in the atmosphere, as we saw atop the oldest Mayan pyramid.

Our ancient mound builders repeatedly sought out places with such natural electrical activities. When you have, as at Cahokia, two large zones of differing magnetic field strength, it normally means that one zone possesses more magnetized iron underground than the other. The ability of the ground to conduct electricity is proportional to its content of water and/or metals. The wetter or more metallic, the better electric current will travel through it.

Cahokia's biggest mound was erected at the edge of a creek. Most of the Hopewell and Mississippian mounds were built near the edge of a flood plain, which would be inundated in spring, just before seeds were planted. Within a floodplain, a zone of magnetic transition will add fuel to the fire, so to speak, as far as maximizing ground currents. The only thing that will do better is a thunderstorm.

A PLACE WORTH DEFENDING

The Cahokians also built a 'Woodhenge' – a 400-foot circle of cedar posts with one post at the center. Sight lines from the center, past certain outlier posts corresponded to sunrise on the two equinoxes and on summer solstice. But this calendrical device was not located anywhere near a magnetic anomaly; magnetism was a connection employed only with the mounds.

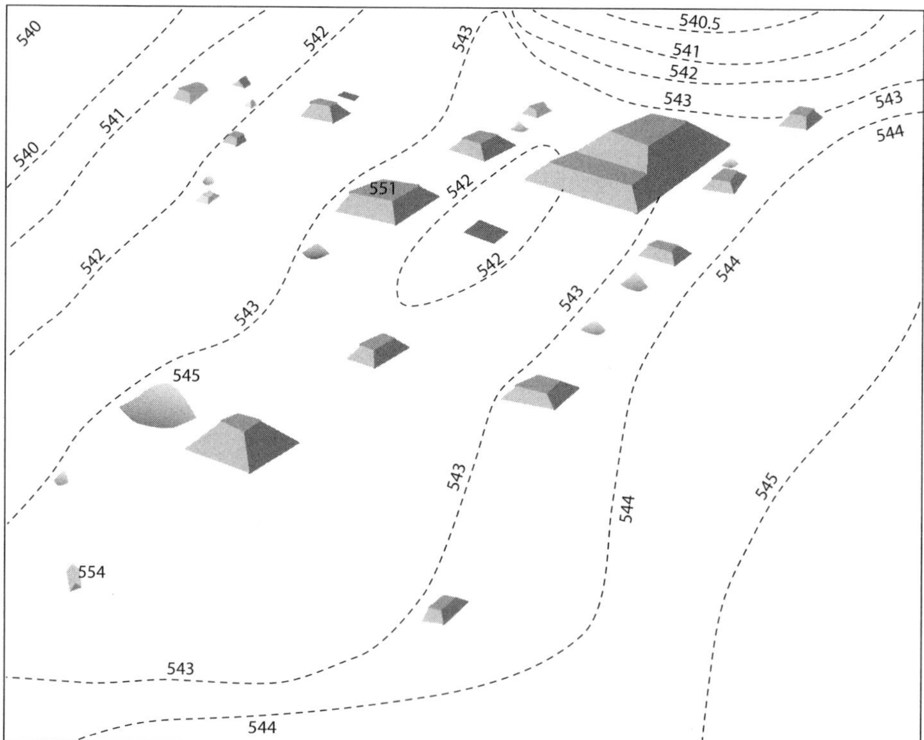

Figure 13. Magnetic contour map of the great mound center at the Native American complex of Cahokia, Illinois. Mounds are placed on positive magnetic anomalies and laid out on axes of 53,300 gammas magnetic field strength. The giant central mound – the biggest in the U.S. – was placed on the edge of greatest transition by the former riverbank. (John Burke)

From Woodhenge (54,000 gammas), on the western edge of Cahokia, the geomagnetic readings rise as you move east. As depicted in Figure 13, every time a reading rose past 54,300 gammas a flat-topped mound would be present – not just once, but repeatedly, in a north-south line. The same is true when you move towards the complex from the north, south, or west.

For a century, investigators have been asking why the Cahokia mounds are on two great axes. In the center of the complex, a north-south line occurs with readings of 54,200 gammas – lower than the 54,300 on either side of it. This is the central axis of Cahokia, with the largest mounds and the great central plaza. Now, obviously not every piece of ground on the 54,300-gamma frontier is covered with a mound. In fact, the mounds were sited where there was also an additional, small, individual magnetic anomaly.

Whatever was occurring here, it made the Cahokians rich. Here also, lavish trade goods flowed in from all over North America. Yet the Cahokians had no special resources to exchange in return. There was something about the mound complex, though, that apparently aroused the envy of their neighbors to such a degree that the residents erected the largest fortifications in the history of pre-Columbian North America to protect it. A palisade twenty feet high, made of trees a solid foot thick, was thrown up all the way around the 2.5 mile perimeter. It did not enclose most of the houses or the rich cornfields. But, it did protect the mounds – and protected them well. Every twenty paces, a tower jutted out so that archers could shoot down on enemies attacking the walls. The logs themselves were covered by a thick coat of dried clay, so that flaming arrows could not ignite them.[14] Apparently, other tribes wanted the mounds badly enough to kill and die for them. When the stockade eventually weakened with decay, the Cahokians would build another, using 20,000 massive trees each time. This happened three times, deforesting the area.

BREATHING LIFE INTO RELICS

Archaeologist Thomas Emerson points out that people in his field are thought of as dealing with the material parts of an ancient people's life, deducing what they ate and how they worked – nothing about the way they thought. He maintains that the best window into the minds of those long gone lies within the symbolism in their art. Cahokians did make beautiful statues and figurines for themselves. In other areas throughout the country at that time, such figurines mostly dealt with warfare. In what Professor Emerson calls a 'highly unusual' exception, Cahokia's art concentrates obsessively on one theme – plant fertility.[15] Two famous female figurines unearthed at Cahokia exemplify this theme. One kneels on ears of corn, while the other bends her farmer's hoe to a serpent which wraps around her and from which vines grow. Authorities agree that this combined motif represents fertility.[16] Were these symbols part of religious rituals at the site?

Cahokia was finally abandoned around 1300 AD, probably due to persistent floods from a permanent increase in rainfall about that time.[17] We will never know in detail what went on at the site – despite the artist's depiction in Plate 22. Fortunately, however, further downriver in Louisiana, the Natchez tribe of the Mississippian Culture continued using mounds right up to the early 1700's, when French colonizers interacted with them. The Jesuit priest Father Le Petit wrote back to his superior in 1730 that a temple of wood stood atop the Natchez's earthen pyramid. No farmer, he reported, would think of planting his fields without first "presenting the seed, with accustomed ceremonies, in the temple."[18, 19]

Excavations have shown that all Mississippian mounds had a thatched roof 'temple' on top. As we shall see later, there are many reasons these mound-top buildings should have experienced the same type of electromagnetic changes as Serpent Mound when a thunderstorm rolled through. However, by having a roof overhead, anyone placing seed in a Mississippian mound-top temple would not have to scramble to remove the seed when downpours arrived. This way, the seed could be left at the place with the most dramatic electromagnetic changes, during the peak period of those energies.

What was the Mississippian 'seed temple' like inside? According to the earliest European visitors to the Natchez who passed through the area with De Soto's expedition in 1540, its outer room contained a constant fire, tended by old men who were executed if they allowed either of two things to happen. If the fire went out, they were dead. They would also be killed if they allowed the simmering, smoking fire to break out into open flame.[20] This 'eternal fire' was to have no flame. No one could offer an answer to the early European explorers asking the reason for this. The tribal priests simply said that it had to be this way.[21]

You can confirm for yourself at your home barbecue that smoke is one large cloud of negatively electrified air particles, or ions. Flame, on the other hand, has very little electrical charge to it. The Mississippians desired only the negatively charged type of fire. Since they deliberately left out any chimney or smoke hole in the roof, these charged particles gathered thickly inside. This outer room of the wood temple was one big repository of negative air ions – the same kind as delivered by thunderstorms.

The inner sanctum was even more intriguing. It was a clay version of the rock chambers, in which we enhanced seeds. Its wattle-and-daub frame of sticks and cane was covered with mud, lined with clay inside, and then covered over in reed mats. It was round and had a domed roof. There were no windows, and the only entrance was through a door four feet high and three feet wide[22], same as the rock chambers. Combined with the effects of the electrified smoke and the magnetic anomaly on which the mounds were sited, this structure created conditions similar to what we found in the rock chambers. It appears that the mound builders exposed their seed to the same seed-enhancing energies as in our rock chambers. We therefore note with interest that the crops on Natchez farms were known for being unbelievably lush and fertile.[23]

CHAMBER MEETS MOUND

In the state of Mississippi, the Choctaw tribe still tells of their ancestral journey from the West to their current homeland. When they ran out of corn to eat on the long march, the shaman made them stop beside an abandoned Mississippian mound, which they called Nanih Waiya, to plant their last seed and start a new stock of corn. According to this highly detailed and credible account, they only had two ears left and these were two years old. Everyone feared the corn was dead and would not sprout when planted. But when they planted it in the ground beside the great mound, it flourished, replenishing their stock of both food and seed corn. They were sufficiently struck by this phenomenon that they stayed there and made the mound the center of their new homeland.[24]

We confirmed that this mound was also built atop a strong magnetic anomaly, spilling beyond the mound in its current form. One muggy summer afternoon, at Nanih Waiya State Park, I found substantial surges in readings during an approaching thunderstorm. Ultimately, I was chased off the mound by bolts that seemed to zero in on it when they began striking.

Of all the improvements in seed that we have seen, the most dramatic changes always seemed to occur with low vigor seed, just what the Choctaw had. This has been true both with seed left at ancient sites as well as that treated with the modern version of these energies.

In Choctaw, Nanih Waiya means 'The Productive Mound.'[25] Another interesting translation is 'to bear' or 'to bring forth.'[26] Though most tribes had to move their farms once or twice a generation due to soil exhaustion, the Choctaw never seemed to suffer from this problem, and their settlements stayed constant over the years.[27] This alone was an exceptional accomplishment. Upstate New York is one of the few places where we have quantitative yield figures for early Native American farmers. Jesuit missionaries living among the Mahican or Mohegan tribe estimated that the tribe's fields yielded twenty-seven bushels per acre.[28] (Early nineteenth century settlers in the same area, who also practiced slash and burn agriculture and planted their corn Indian style, got yields of twenty to thirty bushels per acre. Within eight to twelve years, as soil exhaustion set in, yields would fall to seven to ten bushels per acre, and the entire village would have to move.)[29]

The Mohicans did not have rock chambers. It also bears remembering that these early farmers did not use fertilizer or crop rotation. The myth of Samoset teaching the Pilgrims to put fish heads in the corn holes for fertilizer has never been substantiated by experts in the field.[30] Our own intensive searches through the Library of the Museum of the American Indian produced only some references to the spreading of ground up shells on fields in the Carolinas, which would raise the calcium content of the soil but not its real fertility. In fact, early white observers in Western New York described Indian agricultural practices in great detail without ever mentioning any use of fertilizers.[31] In Virginia, an observer as early as 1587 noted of Indian farmers, "The ground they never fatten with muck or dung or anything . . ." [32]

Somehow, the Choctaw managed to prevent their yields from falling. It is interesting that the Mohegans were not mound builders. This reminds us of the dichotomy observed in Chapter One regarding the Olmec, that villages with a mound enjoyed a higher standard of living than villages without one.

About the time that the great mound center of Cahokia was abandoned, it was replaced by a major Mississippian center in today's Moundville, Alabama. Around 1250 AD, this three-hundred-acre center of multiple flat-topped mounds mushroomed into existence like a planned community. Perhaps the most learned expert on this site, archaeologist Jim Knight of the University of Alabama, Tuscaloosa, says that the mounds symbolized sacred mountains. They were thought of as hollow, representing caves from which, myths say, the first people emerged.[33] This is reminiscent of the beehive-shaped stone 'burial' chambers, lying hidden inside some of the Adena mounds.

For the Mississippians, this death motif was associated with the serpent, as well as with fertility. This was no coincidence. Numerous traditions tell of sacred 'bundles of fertility' and of renewal presented by deities. "It was believed that through the spirits of dead relatives the Creator would help the living . . ." [34] This linkage may help explain the frequent burials at mounds.

Other Mississippian tribes, such as the Cherokee, employed a combination of chamber and mound. Beehive-shaped rock chambers, both with and without skeletons, have been found at the bottom of mounds. Before any mound was erected, a hut like the Mississippian temples was constructed with a fascinating array of contents. Into these huts, corpses were interred, wrapped in bark – just as the dead had been in the earlier Adena mounds. Interestingly, the Cherokee would often germinate seeds in beds of bark before planting. We will see more possible connections between skeletons and seed fertility in Europe.

Next to the Cherokee skeletons were beehive-shaped clay vaults. Located as far north as North Carolina, they were comparable in form and size to the smaller chambers of New England, three to five feet high. In the words of one excavator, "This was partially filled with rotten bark, human bones, and dark decomposed matter." [35] (We would love to know if the decomposed matter included seed.) The builders occasionally crafted semicircles of these small beehive chambers, clustered around what would become the center of the mound. These were a people whose central mythological image was that of 'caves' of fertility inside sacred mountains.

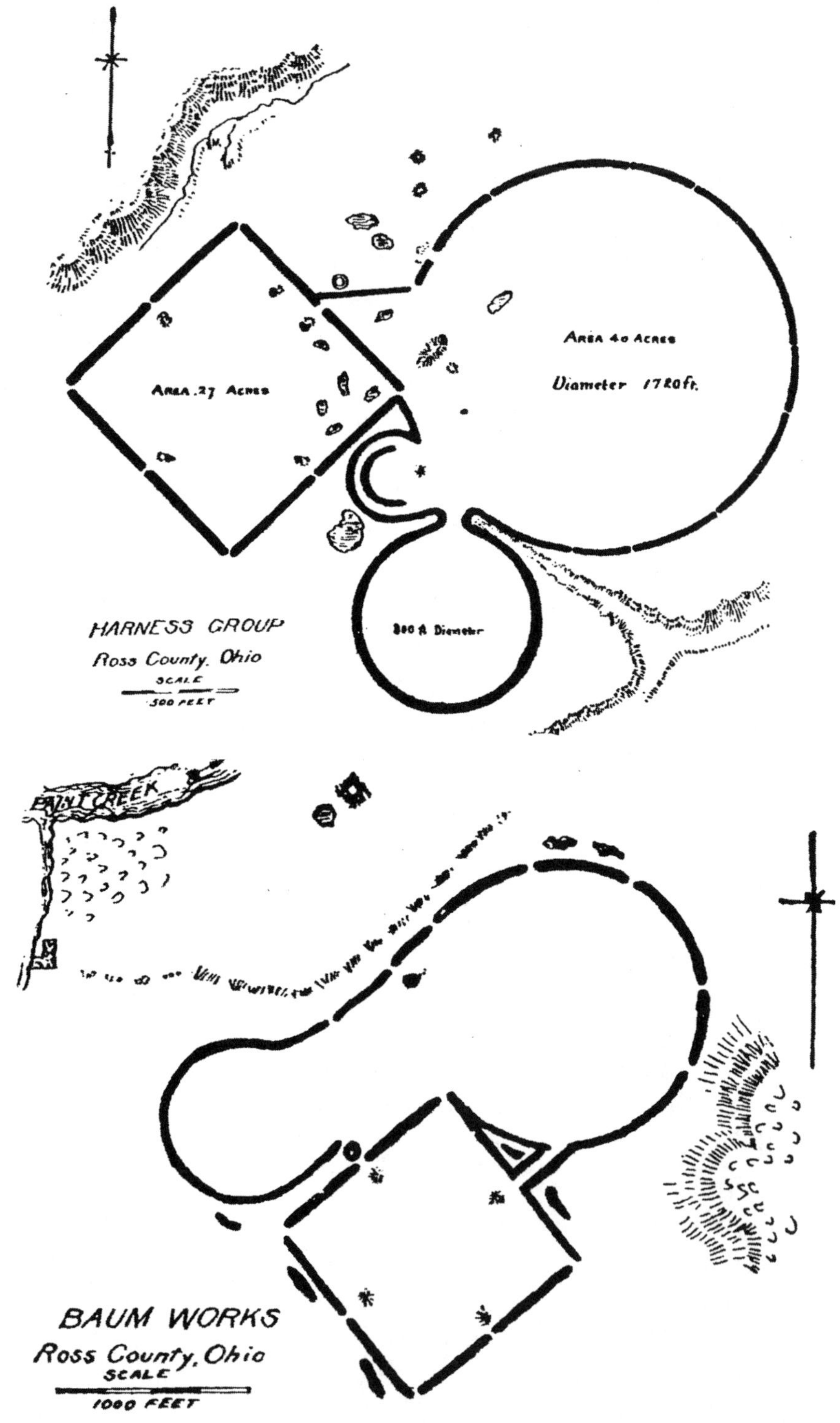

Area 40 Acres
Diameter 1720 ft.
Area 27 Acres
800 ft Diameter
HARNESS GROUP
Ross County, Ohio
SCALE
500 FEET
PAINT CREEK
BAUM WORKS
Ross County, Ohio
SCALE
1000 FEET

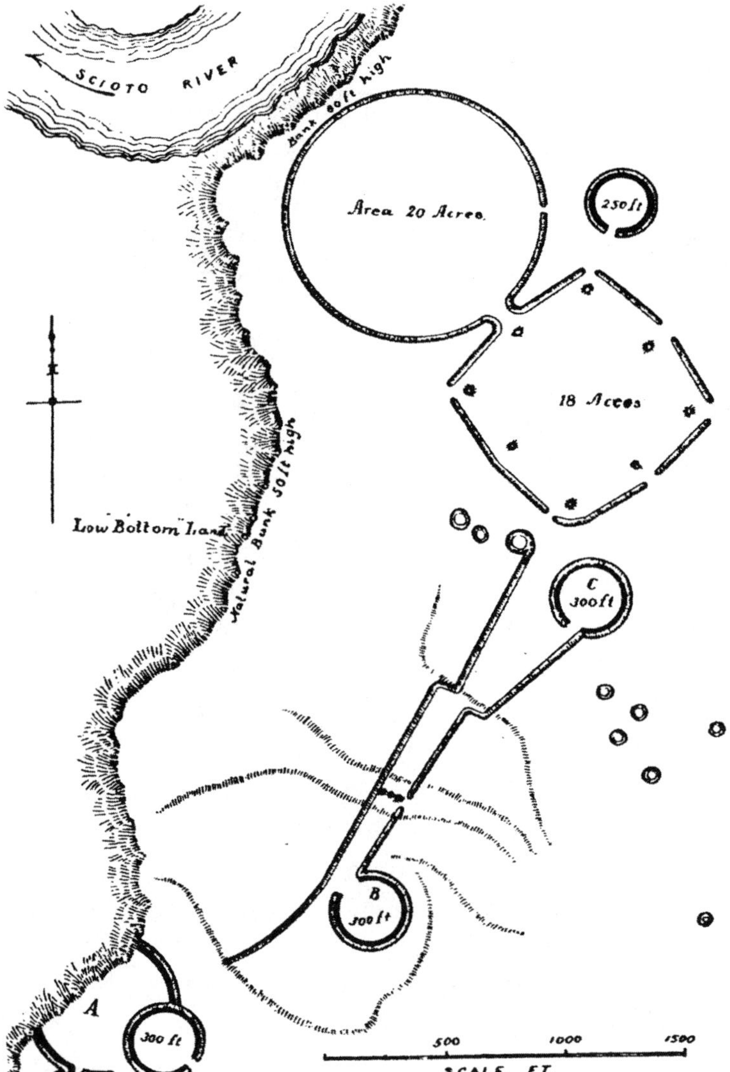

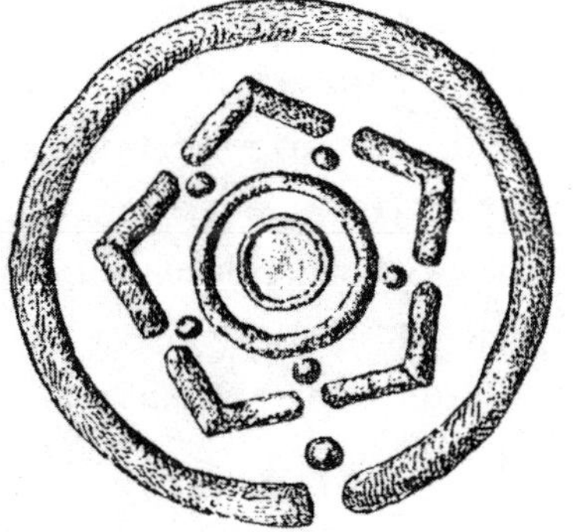

Figure 14. Regardless of their shapes, the American ditch and mound systems commonly have small mounds placed just inside their dicthes' gaps. It seems that the builders wanted to first concentrate surface electrical current in the gaps, or causeways, and then further concentrate them in the mounds. Frequently referred to as 'defensive structures,' they have a serious defensive flaw: Their 'moats' lie *inside* their 'walls.' (Illustration from E.G. Squier & E.H. Davis. *Ancient Monuments of the Mississippi Valley: Comprising the Results of Extensive Original Surveys and Explorations*. Smithsonian Institution, Washington DC, 1848)

In Meso-America, where the corn-centered culture emerged and eventually spread, such caves were considered the 'home of seeds' and the source of fertility. The Mayan glyph for fertility depicts a cave with a beehive inside.[36] Archaeologist Henry Mercer, University of Oklahoma, was brought by his Mayan guides to Oxkutzcab, where they entered an oblong mound by a low door to find a well-preserved rock chamber inside. At twelve by six feet, it was similar in size to the New England chambers. His guides led him on another hundred feet to what they called 'The Cave of the Serpents,' with all its fertility associations.[37]

Aztec and Maya pyramids were sometimes erected over sacred caves. The largest, the Aztec Pyramid of the Sun almost equals the Great Pyramid of Giza in volume. It was placed squarely over a previously used 'oracular' cave that had seen much use and had been artificially re-shaped through tunneling and wall-building. The cave itself was a portion of an extinct flow tube of lava – a highly magnetic mineral. It had been artificially altered into a four-leaf clover, or quatrefoil shape – the seat of fertility. Mayan art frequently depicts corn foliage emerging from a quatrefoil cave.[38] Quatrefoil chambers inside mounds are found throughout the Old World as well; for example, at Newgrange in Ireland. In the words of archaeologist Doris Heyden, "The cave or grotto is the place where sustenance is kept, it is the depository of seeds . . ."[39] Items buried within the Pyramid of the Sun, dedicated to Tlaloc, the fertility god, included a complete skeleton of a jaguar, long the icon of plant fertility throughout Meso-America.[40, 41]

One Aztec emperor, after a vicious famine in 1250 AD, launched a project, which required reshaping an entire hill, called Texcotzingo. At the foot of this hill was an artificial cave. Early Spanish chroniclers reported that every spring priestesses of the maize goddess (named 'Seven Serpent') would bring the best seed corn from last year's harvest up Texcotzingo to the temple of the maize goddess and would leave it there long enough for certain blessings. Despite the fact that most of their fields were left fallow, or unplanted, in any given year, these sixteenth century Aztec farmers grew so much corn that they comfortably supported a population density greater than that of Western Europe at the time.[42] Such caves are, in fact, still considered by modern Aztec descendants, the Nahuatl, to be the natural home of seeds.[43]

EARTH MEETS SKY

The Hopewell not only built mounds, they also dug ditches. Since telluric currents travel primarily in the top few feet of ground, digging even a modest ditch breaks their flow, like a fallen tree in a shallow stream. The current will try to flow around the obstacle, seeking the path of least resistance. The Hopewell would leave certain patches of ground intact, creating breaks or 'causeways' that would allow all the current blocked by the ditch to rush through, directing the current flowing through a wide area into a tight bottleneck of intact ground. Many acres were enclosed this way, like the flood plain of Paint Creek near Chillicothe, Ohio. Today, most ditches have disappeared, leveled by farmers; others are still visible at places like Hopewell Culture National Historical Park.

Just inside these bottlenecks, the Hopewell often placed mounds, further concentrating the telluric currents and 'charging' the mounds (Figure 14). As we shall see in Chapter Nine, a similar technique was used in England five thousand years ago, to construct causewayed enclosures and henges. This concentration of telluric currents at these particular points could create a node or duct, bridging the natural electromagnetic energies of the atmosphere and the land. Thus, American mound builders were connecting the energies of Mother Earth and Father Sky. 'Ceremonial centers' is the term archaeologists have applied to these groups of mounds, and we see no reason to argue with this name. We must remember, however, that our modern idea of 'ceremonial' means 'no physical effect' of any consequence. This need not have been true in the minds of the ancients.

In his book, *Indians of North America,* Harold Driver writes, "The distinction between natural and supernatural was never sharply drawn by Indians, who tended to blend the two into a harmonious whole . . . a certain amount of practical science often went along with the supplication of the supernatural. For instance, Indian farmers everywhere combined practical science with religion and magic. One without the other was inconceivable." [44] Authority Ake Hultkrantz boldly states that "The chief role of American Indian religion was to help the crops grow better." [45] And so we see, everywhere, connections between earth and sky, between the New World and the Old, between the physical and the ceremonial.

CHAPTER 8

The First Megaliths

"It is amazing – people lived in simple dwellings of wood that are long gone; they made a huge effort in stone, in the same places, for thousands of years – why?"

– Pierre Mereaux[1]

In an attempt to answer the above question, the late Pierre Mereaux spent every vacation for thirty years surveying the electromagnetic properties of the world's oldest stone creations. An unfathomable 6,700 years ago, inhabitants of one corner of France began raising enormous standing stones and rock chambers. Two thousand years before Egypt saw its first pyramid, these French 'engineers' had zeroed in on energetic ground and seemed to be harnessing it.

Mereaux was the right person to investigate. Trained as an engineer in applied thermodynamics, his technical skills and persistent, organized approach were the best means to make sense of a small region that contains 11,000 standing stones and dozens of chambers. It would take a greater challenge than this to intimidate a man who had spent World War II blowing up bridges for British Intelligence in German-occupied Belgium.

Mereaux was convinced that more was going on in ancient France than met the eye – something that involved invisible forces, detectable by the instruments he brought to bear year after year. But let us start at the beginning.

FEAST OR FAMINE

The simple model of the spread of agriculture that so many of us were taught in school is not accurate. Although farming did originate in the

Middle East about 8500 BC, it did not then spread quickly westward to Europe and revolutionize the world. For two thousand years or more, farming remained primarily a part-time activity, practiced by bands of hunter-gatherers.[2] No one was anxious to give up the nature-based traditional lifestyle. Far from the 'feast or famine' image we were taught, nomadic hunters had varied diets and lived longer, healthier lives than sedentary farmers, who ate a monotonous diet and were, therefore, riddled with deficiency diseases. Full-time agriculture was eventually embraced as a way of life in Europe but on a spotty basis, and it was not driven by envy of the farming life, but rather by sheer desperation.

In the early days of Stone Age Europe, many hunter-gatherer tribes lived near the coast. Here, particularly in estuaries, food was abundant: fish, fat seals, seabirds breeding in large colonies where eggs were plentiful, crabs, and shellfish. Oysters in particular, rich in calories and tasty, formed an important mainstay of the diet for these people. Being able to feed many children, these bands flourished, and populations became dense around these estuaries and coastal marshes. The formerly nomadic people became sedentary. Life was good.

Then a major change in climate during the fifth millennium BC produced a major increase in rainfall, changing the salinity of the estuaries. This killed whole populations of oysters, which can only reproduce in narrow ranges of salinity. And just like that, a major food source vanished, taking with it a brief Golden Age.

What had been a comfortable population density now became deadly. Too many people had lived stationary lives for many generations, thanks to the abundance of shellfish. The other natural food sources of the sea were not enough to feed the population, and returning to a normal hunter-gatherer economy inland was simply not an option. Over time, the inland territories had become occupied by other bands. There was simply no new land to which to move. These people literally had their backs to the sea, and their situation was desperate. Lacking any real choice, they turned to the only alternative available – full-time farming.[3] Only in this way could they feed their dense population on a relatively small parcel of land.

	Agriculture appears	Dolmens/ rock chambers appear
Iberian peninsula	c. 5500 BC	c. 4400 BC
Southwestern France	before 5000 BC	c. 4600 BC
Denmark	c. 4200 BC	c. 3600 BC

Table 1. First appearance of agriculture, compared with first appearance of dolmens or rock chambers at some European locations.

THE GIANT *MENHIRS*

The Gulf of Mourbihan in France's Brittany was one such place. With the Atlantic to the west, the long narrow peninsula of Locmariaquer protects to the east an extensive estuary, which was rich in oysters until the climate change occurred. The soil was fertile at first. However, practicing the ancient slash-and-burn agriculture quickly exhausted the soil, and within a few centuries it would become difficult to feed the population on the limited amount of land available. As early as 1976, Colin Renfrew argued that the megaliths were built in response to this pressure.

Around 4700 BC, these farmers began to build giant stone structures.[4] As we see time and again, the building of these structures didn't happen at just any time. In each location it occurred only after there had been full-time farming for centuries. Using the slash-and-burn agriculture of the era, the soil quickly becomes exhausted and food production falters. Table 1 shows the sequence of these events in different parts of Europe.[5]

Megalith means 'large stone,' and France was the place where the 'Megalithic Age' began. In fact, when measured by sheer volume and tonnage, far more megalith building occured here than anywhere else in Europe. And from then on, the society of Carnac, as this place is today known, flourished. Its chronology here has long been established. However, in recent decades, two highly unusual individuals (one Belgian, one English) have revolutionized our understanding of the likely purpose of these, the world's very first megaliths.

The earliest creations were single tall stones, which we call by the French term *menhir.* They were erected at certain sites and frequently decorated with agricultural images such as oxen, plows, and the stone axes that Neolithic farmers employed to cut down trees for fields. Some of the menhirs were immense. The champion, Le Gran Menhir Brise, which is now in four pieces, stood an awe-inspiring sixty-seven feet and weighed over eighty tons. It has been estimated that it took a minimum of 3,800 adults to hoist it into place, and this would have meant the cooperation of numerous separate communities (Figure 15). An entire complex of such decorated menhirs was erected on the Locmariaquer peninsula.[6]

It has been assumed that the menhirs were ceremonial, symbolic structures, probably with religious associations. At Le Manio, a decorated menhir had numerous ceremonial axes deposited at its base, suggesting a link between ax exchange and megalithic ritual from the earliest stage.[7] This link of megaliths and stone axes became endemic in following years, and stone axes from all over France began streaming into the area.

Mereaux, however, believed that the axes were used as symbols of fertility. He also said that the menhirs are known to attract lightning. This is not surprising because at the base of these giant stones he measured negatively charged electric ground currents that interacted with the positive atmospheric electric field. Again, we see a familiar pattern of natural energies, combined with fertility.

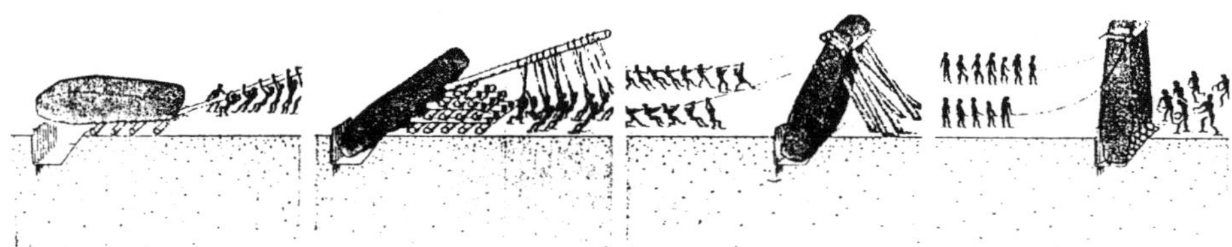

Figure 15. At Carnac, many people were needed to drag, slide, pry, and hoist the giant stone *menhirs* into place. The 67-foot Grand Menhir Brise, weighing over 80 tons, would have required about 3,800 adults. [From R.J.C. Atkinson, *Stonehenge and Neighbouring Monuments,* English Heritage, London, 1987, p. 15, by permission.]

ARE PASSAGE GRAVES REALLY GRAVES?

The next style of stone structures to arise in Brittany was the so-called passage graves. They are like larger versions of our New England rock chambers of Chapter Six, with additions. And, as with their American cousins,

Mereaux found, they were placed on spots of magnetic variation. Entrance was gained through a small opening, leading to the central chamber via a long, narrow tunnel that was lined and roofed with stone slabs. After completion, the entire assemblage was buried under an enormous mound. Carbon dating has placed the origin of passage graves here at the astonishingly early date of about 4700 BC.[8]

Skeletons have been found in these passage graves, and an individual grave would be used for several centuries. Yet in all that time, only two or three dozen individuals might be buried there. In other cases, such as the largest of all passage graves – Newgrange in Ireland – only five or six burials took place. An entire region might contain just a few of these passage graves.

This scarcity has caused some archaeologists to argue that the primary use of 'passage graves' was not at all as graves. Rather, the burial aspect seems to have been a ceremonial part of some more important function. What bones are found were usually brought there and burned. English expert Colin Burgess explains, "It is strange that no tomb was ever filled with human remains to anything like its capacity, not even the Irish passage graves where considerable accumulations of burnt bone are common. It is likely that the burials at these sites were part of a much wider range of ceremonial and ritual functions, and that they were certainly not intended to act as charnel houses for successive burials."[9] Ian Kinnes of the British Museum in London adds, "It was always rather blithely assumed that their function was for burial. Now there's an increasing feeling that this may be only one aspect, and perhaps in some instances only a minor aspect, of very complex rituals which reflect the preoccupations of an early farming society facing a new environment."[10]

As we have seen in the Americas, skeletons and skulls were often associated with regeneration and rebirth. In Mexico, for example, human bones were "regarded as the seat of the essential life force and the metaphorical seed from which the individual, whether human, animal, or plant, is reborn . . ."[11] In certain parts of Africa, a planter would dig up an ancestor and bring his skeletal hand to the field during planting season. As late as the early twentieth century, farmers in Finland would borrow bones from the cemetery to place along the edges of their field while plowing, then return them later in the year. In Hungary, dirt from a fresh grave sprinkled on your field was always considered beneficial. The linkage of bones, the dead, and agricultural fertility is an extremely long-standing tradition throughout the World. In Anatolia, Turkey, at Catal Huyuk, the world's first farming culture circa 6,200 BC, the art is dominated by combined images of death and fertility.[12]

CEREMONY OR COMMERCE?

The rock-enclosed, dry-wall, corbel-vaulted chamber evolved in France over the next centuries. The interior is smaller than the vaults of the preceding passage graves. In fact, they are identical in pattern to the New World rock chambers. These were the ones that would dramatically enhance the performance of seeds placed inside, a process which would be of tremendous economic importance to a farming community with limited land and exhausted soil, as was the case at Carnac.

Pierre Mereaux brought his own magnetic gradiometer and ground electrodes to bear on a dozen passage graves and rock chambers. Setting the magnetometer on its tripod atop the capstones, he found that the needle of the meter often 'danced all day' on a chamber, but remained quite steady in the adjacent fields. His magnetometer worked a bit differently than ours. It would register changes in the strength of the vertical component of the magnetic field, but not absolute strength. Ours measures absolute strength, but is not as sensitive as Mereaux's. Yet, time and again, his readings coincide with what we have obtained at numerous locations. Mereaux also measured the electric ground currents inside and outside chambers. The current was always many times stronger inside.

Mereaux is not the only revolutionary to shed new light on archaeological assumptions regarding Carnac. In the past twenty years, in fact, a mutiny of sorts has begun within the conservative confines of French archaeology. Cast in the role of leader of these mutineers is an Englishman: Dr. Mark Patton. As curator of the archaeological museum on the Island of Jersey, he has spent nearly two decades excavating these structures, and some of his finds have surprised the academic world.

Formerly, it was always assumed that the huge menhirs were erected for ceremonial, even religious, purposes. Thus, one would expect them to be accorded great respect. In 1984, however, while excavating the geometrically decorated megalithic site at Gavrinis, French archaeologist

Charles-Tanguy Le Roux discovered that the carvings on one of the giant stone slabs used in the ceiling exactly matched those on the capstone of Locmariaquer's famous Grand Tumulus La Table des Marchands. But as subsequent investigation revealed, even the Grand Tumulus was not the original source of this stone. Both slabs had come from the same original block that once stood in the ground as a decorated menhir. More recent excavations provide clear evidence that earlier menhirs, and even the more recent Grand Tumulus structures, were regularly uprooted, broken up, and re-used in later megalithic structures. [13]

As we noted earlier, the peoples of the Americas blended the spiritual, ceremonial, and practical, especially in the vital area of fertility. If the later structures had a primarily practical value, with each model being superseded by a better model, it makes very good sense indeed to reuse material from the older structures. It would not be unlike stripping boards off an abandoned building to erect a new one.

When arguing for a strictly religious or ceremonial purpose behind the massive effort needed to build these structures, people often use the comparison of the volunteer craftsmen who erected the European cathedrals in the Middle Ages. These were people moved by religious passion, they say, and they required no pay for their labors. However, this argument ignores important differences. The craftsmen of Paris who erected Notre Dame, constituted less than one percent of the population of the city. Many, perhaps most, were, in fact, paid. The volunteers probably worked on the cathedral in their spare time. By contrast, many great megalithic structures had to involve twenty percent of the population, many of them full time, for years, or part time for decades – even centuries. Neolithic society made far more of an investment per capita in building the ancient structures than did any Medieval European city in creating even the most impressive cathedrals. Furthermore, the Neolithic structures were usually built at a time of desperation. The Notre Dame's of the world were products of affluence that could spare whole segments of the population for non-food related activity. The situations are not comparable.

Patton provides impressive evidence that the changes in megaliths were part of a change in Neolithic social structure. He believes that the tribal elders in France were "replaced by a new elite, whose status depended more on accumulated wealth than on ancestral lineage." [14] If so, where did the wealth come from, and what was the connection between wealth and megalith building?

AN INQUISITIVE BELGIAN IN BRITTANY

Brittany is the westernmost part of France and is mainly rocky. Geologically, it is a part of the Armorican Massif. Rich in granite, quartz, and schist, it is also loaded with magnetite – the magnetic rock so intimately associated with the North American rock chambers. A part of Le Mourbihan peninsula has a good amount of magnetite in the ground and, accordingly, is littered with magnetic anomalies. The readings we have discussed in previous chapters average 300 to 500 gamma differences from their surroundings. At Carnac, Mereaux's readings ranged from minus 400 gammas to plus 1,100 gammas.

The Locmariaquer peninsula is also the most seismically active region of France, surrounded by thirty-one faults. Figure 16 shows that at least four of the major megaliths lie in a straight line precisely on top of one of the invisible faults.[15] Seismic stresses can cause electrical currents in the ground by two different mechanisms. The most familiar is called the 'piezo-electric effect' and is limited to quartz. It has long been known that quartz placed under pressure develops electric charge, and this area of France is plentiful in quartz.

A second mechanism has been discovered in recent years. Almost any kind of rock, if placed under enough pressure to fracture will emit electric current just before fracturing. This may be the primary cause of newly discovered electric and magnetic signals preceding earthquakes.[16] The mysterious balls of light rising out of the ground, which frequently precede quakes, have been thought to arise from the crushing of rock under stress. In labs, this breakage has produced similar light balls, even in non-quartz rocks.[17] In a prelude to an earthquake in Quebec, basketball size globes of light were seen emerging right through the asphalt of a parking lot.

Armorican geology also produces numerous variances in the strength of the local gravitational field near Carnac. Figure 17 shows how the major stone structures built here align with these zones of disturbance.[18] The three dotted lines show the borders of the areas disturbed in these ways. Outside the lines, readings are uniform. Inside, they vary widely. There is one frontier where the edges of all three zones come together just above the alignments numbered five, six, seven, and eight. And right here, where Mereaux found that the edges of the magnetically, seismically, and gravitationally disturbed zones meet, the ancients erected the most fantastic stone rows in the world.

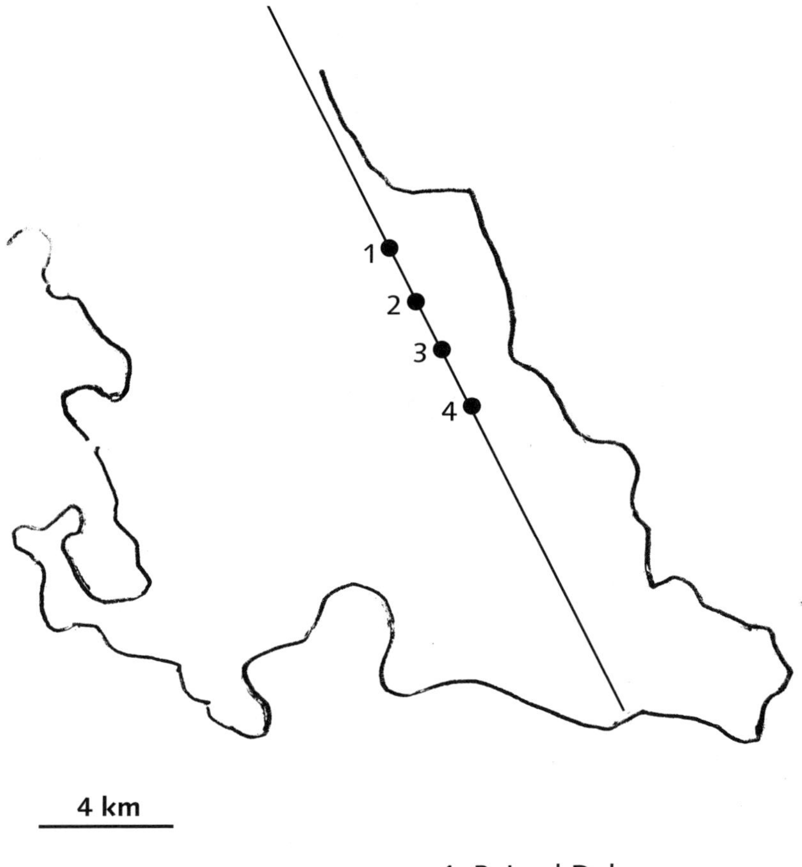

1. Ruined Dolmen
2. Mane-Lud
3. Table des Marchand
4. Mane-Rutual

Figure 16. The major stone alignments at Carnac are generally placed where an overlap occurs on the border of a disturbed zone of magnetic, seismic, and gravitational forces. Such overlaps are extremely rare, and the stone rows here constitute the greatest megalithic structures in the Old World north of Egypt. The builders grew rich. (After Pierre Mereaux. *Carnac – une porte vers l'inconnu.* Editions Robert Laffont, Paris, 1981).

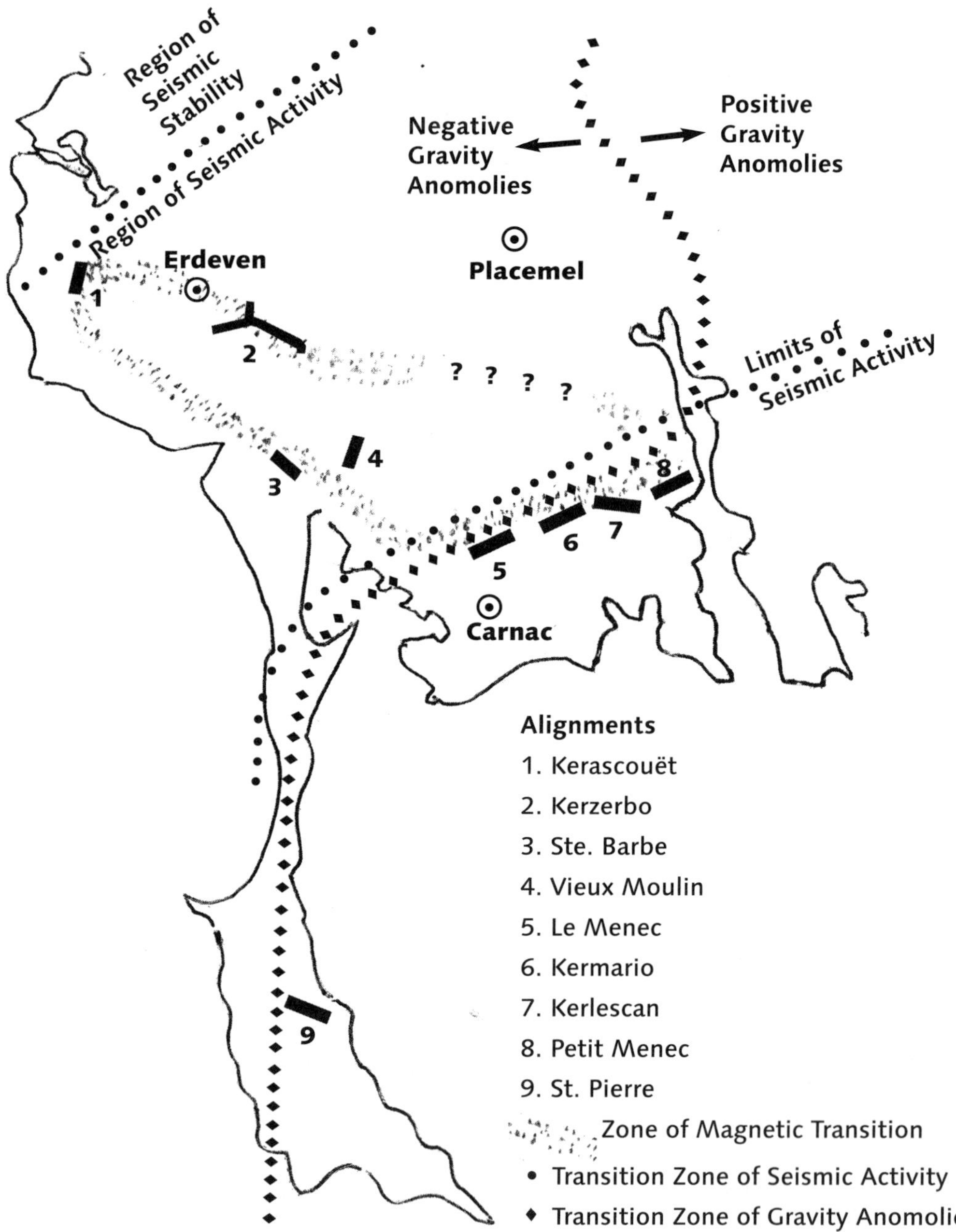

Figure 17. Over the centuries, at the tip of the Locmariaquer Peninsula near Carnac, megaliths were built directly in line, right above a hidden fault. (After: Pierre Mereaux. *Carnac – des pierres pour les vivants.* Nature et Bretagne, Spezet, Bretagne, 1992, p. 59).

When dealing with electromagnetic forces of the earth, experience shows that the strongest effects usually occur at the edge of the disturbed zone (not at its center). On a similar border, the ancient North American metropolis of Cahokia was built. The discontinuity of a gradient seems more important than being where the forces are strongest. At the edge of such discontinuities in electrical conductivity, extreme instabilities in the vertical component of the local geomagnetic field occur,[19] inducing electric currents. Mereaux measured both simultaneously at Carnac.

'MEGA MEGALITHS'

The Grand Tumulus structures were the climactic stage in the erection of megaliths in France. This was an ambitious, though brief, era. The tumuli are even larger than the already massive passage graves, the smallest tumulus being similar in size to the biggest passage grave (about 32,000 square feet, or the size of a football field). In all Grand Tumulus structures, blocks of granite were piled atop one another to massive proportions, sometimes to the height of a five-story building. Table 2 shows the dimensions of six tumuli.[20]

Tumulus	Length	Width	Height
Er Grah	400 ft.	160 ft.	?
Mane-Lud	260 ft.	165 ft.	20 ft.
Manever-Hroek	330 ft.	200 ft.	20 ft.
Le Moustoir	290 ft.	130 ft.	26 ft.
Tumiac	180 ft.	180 ft.	50 ft.
St. Michel	410 ft.	200 ft.	40 ft.

Table 2. Dimensions of six Grand Tumulus structures at Carnac, France.

Enormous though they were, these structures covered chambers no bigger than your average modern bedroom – seven to thirteen feet long by three to ten feet wide and perhaps seven feet tall (the same height as American rock chambers). They were not corbelled, always rectangular in shape, and would sometimes have subsidiary structures such as side niches and cysts in the floor. With the exception of the St. Michel Tumulus, there was only one chamber in each.

They were not built as graves. At St. Michel, twenty-one stone 'burial chambers' were found, but in many only the skeletons of young girls were interred – the favorite sacrifice of planting societies. Other chambers contained only skeletons of small cows.[21] Perhaps the megaliths here were not erected for corpses, but rather the corpses were interred for the megaliths.

So what was their primary purpose? When we look for clues to this puzzle, we see that the tumuli were always made of granite, having a magnetite content ranging from four percent to thirty percent.[22] Granite is the most common radioactive mineral, though, of course, the radioactivity is extremely low level. But walk into any hardware store in New Hampshire, the Granite State, and the home radon detector is always prominently displayed. Every house made of granite needs one. Radon gas is radioactive and constantly emitted by granite. Actual radiated neutrons are also produced by this common mineral. If you live in such an area, you are supposed to always keep a window slightly open. As long as air circulates you will be safe. But when the indoor air is so enclosed as to no longer circulate, radon gas can build up to levels that are believed to produce cancer over time. Geiger counters chatter away in such homes.

With the Grand Tumulus design, we have a small mountain of granite enclosing particularly small air spaces. The airspace has no windows and is only connected to the outside atmosphere by a long tunnel. This design practically eliminates air circulation within. Radiation and radon gas both do one thing extremely well – create ions. The air within the chamber of a Grand Tumulus should be highly ionized on all but the windiest of days.

In addition, the usual energies selected in the siting of passage graves were also sought out. As shown in Figure 16, Mane-Lud lies astride a hidden fault near the tip of the Locmariaquer peninsula, the home of the original menhirs. The tumuli St. Michel and Le Moustoir both flank the multiple-energy frontier zone, on which the alignments lie.[23] A Grand Tumulus should have acted like a giant, particularly intense version of our New England rock chambers. So, quite possibly, these tumuli were used as giant seed treatment chambers.

STONE ROWS

The higher degree of electrification of the air inside Grand Tumulus chambers may have reduced the length of time seeds needed to be left there. This would speed up the amount of seed handled. In modern treatments used in the seed industry, higher voltages mean that only seconds are required to increase seed productivity rather than the many minutes needed in the New England rock chambers.

Stone axes found in the area indicate that people were coming from all over Western Europe to patronize these structures. So demand may well have been outstripping supply. Seed treatment on a massive scale, though, may have been made possible with the arrival of the stone rows.

Containing 11,000 rocks, the stone rows on the Mourbihan Peninsula were the most massive of these rows ever erected anywhere on the planet. The stones, weighing several tons apiece, were stuck upright in the ground to form long lines that would often converge on stone semicircles[24] (Figure 17). Just like at Avebury in England, the menhirs alternate between a lozenge-shaped 'female' stone and a pillar-like 'male' stone.[25] Plate 23 shows an example of these rows.

Going from east to west along the boundary of the zone of disturbance, the Field of Menec runs east by northeast for 3,828 feet, with an average width of 330 feet, the length of ten football fields and twice as wide! Here, eleven lines contain 1,099 menhirs mounted vertically in the ground. At the end of these lines, there seems to have originally been a half circle of standing stones, called a *cromlech.* The stones begin at about two feet in height and continually increase in size towards the cromlech, topping out at thirteen feet tall. This pattern is also found at Avebury Henge in England and, as we shall see in Chapter Nine, there is a very good physical reason for doing it.

The Field of Kermario begins 1,100 feet after the Field of Menec ends. It runs for 3,674 feet and averages 330 feet in width. 982 menhirs are placed in 10 lines, ranging from 1.6 feet to 21 feet high. Here, too, the stones gradually increase in size toward the end.

The Field of Kerlescan begins 1,300 feet later, has a length of 2,886 feet, and an average width of 456 feet. Its 13 lines run due east, totaling 540 stones that range from 2.6 feet to 13 feet high. In each of these three fields, the stone rows are only approximately parallel, but converge gradually as they approach the cromlech.[26]

One cannot escape the striking similarities in scale and numbers of these three fields. They become all the more thought provoking when you look at Figure 17, where all three fields parallel the remarkable convergence of seismic, gravitational, and magnetic frontiers. As we are about to see, the physics of these stone rows is indisputable. In fact, you can test it yourself in a small backyard experiment. Twentieth century physicists used the same principles to build enormous structures, called super-colliders. (More on this in the next chapter.)

Plate 1. *Limestone Mayan pyramids were called 'maize mountains' by their builders and often were situated by a cave, believed to possess powers of fertility. Even today, Mayan farmers emerge from the rain forest at dawn, place bean and corn seed atop the pyramid, and then take them back home. These Mayan ruins are the so-called King's and Queen's Pyramids at Tikal, Guatemala. (Photo © Kaj Halberg)*

Plate 2. *This highly energetic, flat-topped pyramid, known as The Lost World Pyramid of Tikal, was the first stone structure built here in the first of Mayan cities. It provides a link with the Mayans' predecessors, the Olmec. In continuous use for 1,300 years — from 600 BC to 700 AD — it is situated on ground that contains extremely powerful electromagnetic energies. The structure itself further concentrates these energies, and readings obtained here by the authors one cloudless morning exceeded those obtained in thunderstorms. (Photo © Kaj Halberg)*

Plate 3. *View down the steps of the King's Pyramid, Tikal in pre-dawn darkness. The flash shows only one small light ball, and our instruments showed no unusual electric charge present. As shown in Plate 13, growth of corn seed placed here was not improved. (Photo © John Burke)*

Plate 4. *View down the steps of the Lost World Pyramid, Tikal at 5:10 A.M. The flash shows many light balls, and our instruments showed extremely powerful electrical activity. As shown in Plate 15, corn seed placed here this morning showed dramatic improvement in growth. (Photo © John Burke)*

Plate 5. *View across top of the Lost World Pyramid, Tikal at 5:10 A.M., during a period of extremely powerful electric charge in the air. The flash photo is choked with overlapping light balls. As shown in Plate 15, corn seed placed here this morning showed enormous improvement in growth. (Photo © John Burke)*

Plate 6. *View down the steps of Lost World Pyramid, Tikal at 7:40 A.M. The flash shows no light balls. Our instruments showed that earlier electric charge had dissipated by now. (Photo © John Burke)*

Plate 7. *Top of the King's Pyramid, Tikal, photographed at 6:45 A.M. Our instruments showed an absence of any unusual electric charge in the air. The flash shows only one small ball of light. As shown in Plate 13, growth of corn seed placed here was not improved. (Photo © John Burke)*

Plate 8. *View across top of Lost World Pyramid, Tikal at 7:40 A.M. The flash photo shows team member Geoff Groom standing amid a moderate amount of light balls. The instrument showed that earlier intense airborne electric charge had weakened a great deal, but still remained strong here at the pyramid's top. In the foreground, instruments are recording the electrical activity. In the midground, different types of corn are placed on the pyramid. (Photo © John Burke)*

Plate 9. *Co-author Burke standing at the densest cloth cluster of all on Bear Butte, twin festooned trees framing a stunning panorama of the plains to the north at the edge of a thousand-foot cliff. Right here was the most powerful magnetic anomaly we have ever found at an incredible 900 gammas. The shaman who decorated these sites marked every single magnetic anomaly that our magnetometer measured, showing he was able to detect such forces by sheer physical sensitivity. (Photo © Kaj Halberg)*

Plate 10. *At the top of Bear Butte, a spot had been marked with small piles of rocks holding two-foot willow 'flagpoles' that marked the cardinal directions. Connecting the four flags were strings covered with hundreds of knotted bits of brightly colored cloth. This spot, at 500 gammas the strongest of the five ridge-top anomalies, is a Native American site for vision quests. (Photo © Kaj Halberg)*

Plate 11. *On Bear Butte, the great Sioux leader Crazy Horse was taken on vision quests by his spiritual mentor and childhood friend, who invented the system of colored flags and willow sticks still used today. In the meadow by today's parking lot at Bear Butte, the Sioux would gather in the morning to wait for Crazy Horse to descend from the butte and preach his visions. (Photo © Kaj Halberg)*

Plate 12. *Seedlings of contemporary Mayan corn, eleven days old, germinated as a control sample kept in our hotel room at Tikal. (Photo © John Burke)*

Plate 13. *Seedlings of same type of Mayan corn as above, eleven days old, germinated after being placed on top of the King's Pyramid, Tikal on a morning with no appreciable airborne electrical activity. The seeds were in fact doing worse than the control sample. The reason could be that these seeds were left out on a very humid morning, which may have harmed them. (Photo © John Burke)*

Plate 14. *Seedlings of same type of Mayan corn as in Plates 12 and 13, eleven days old, germinated after being placed on top of the Lost World Pyramid, Tikal on a morning with moderate airborne electric charge. (Photo © John Burke)*

Plate 15. *Seedlings of same type of Mayan corn as above, eleven days old, germinated after being placed on top of the Lost World Pyramid, Tikal on a morning with extremely powerful electrical activity. The seeds show dramatic improvement in growth. (Photo © John Burke)*

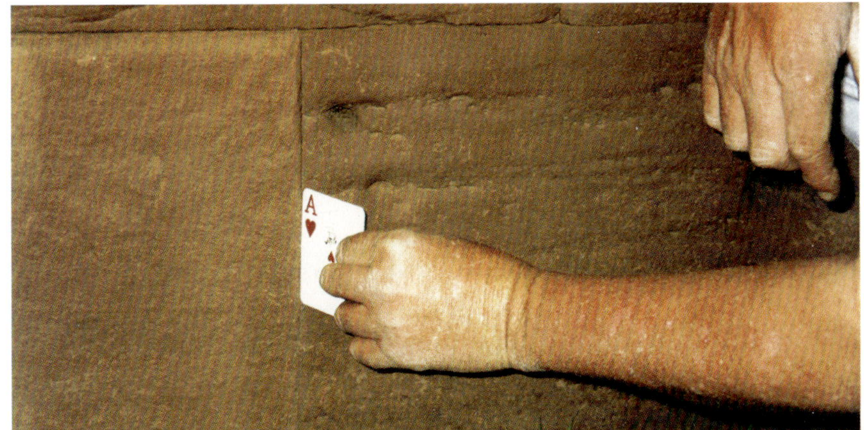

A.

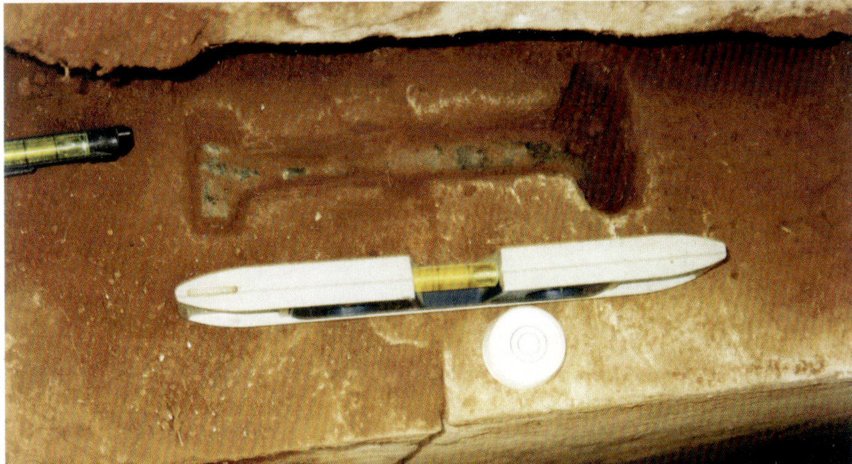

B.

C.

Plate 16. *The electrically conductive stones of the Akapana pyramid in Tiwanaku were so precisely cut and fitted together that even today a playing card cannot be inserted between them (A). The stones were connected by highly conductive copper clamps (B). A reservoir atop the pyramid held water that could be released by a sluice gate to cascade through special drains (C) and generate electric charge throughout the pyramid. The empire grew wealthy from vast agricultural surpluses at an elevation of 12,500 feet, where little grows today. (Photos © Carol Cote)*

Plates 17 and 18. *Putnam County, New York, is rich in magnetite ore and in corbel-roofed rock chambers. The possible plasma arch in Plate 18 at the left side of the entrance 'fried' an electrostatic voltmeter left in an outer pocket of the backpack seen in Plate 17. The air inside this chamber is unusually electrically stratified. Corn, wheat, and bean seeds placed in the chamber repeatedly showed statistically significant increases in germination and growth. This magnetite-rich area has dozens of such chambers, usually located near water and directly on a magnetic anomaly. (Photos © John Burke)*

Plate 19. *This infrared photo, taken inside a Native American rock chamber, shows a floating ball of energy that was invisible to the naked eye. The large stone roof slabs directly above it are highly magnetic. The authors left seeds of 'Iroquois Blue Flint' corn in this chamber for seventy-five minutes, directly under the spot where the glowing ball of energy appeared in this prior photo. The seed — planted and grown with traditional ancient Native American farming methods — grew almost 100 percent faster than a control sample. (Photo © Anastasia Wietrzychowski)*

Plate 20. *Top row of 'Iroquois Blue Flint' corn was harvested from forty seeds, placed a hundred feet outside the chamber shown above. Bottom two rows were harvested from forty seeds, left inside the chamber for seventy-five minutes. The chamber seeds produced three times as much corn by weight. (Photo © John Burke)*

Plate 21. *Balanced Rock, a boulder weighing over 60 tons, perched precariously on three flat limestone slabs, lies at Route 116 in North Salem, New York. Geologists have labeled it an accidental glacial erratic, but it sits precisely over a major magnetic anomaly, and brilliant golf-ball sized lights have repeatedly been photographed circling it. In this flash photo, a few fainter light balls are seen. (Photo © Roberta Puhalski)*

Plate 22. *Artist's re-creation of the fertility-oriented Native American metropolis at Cahokia, Illinois. The Mississippian Culture erected colossal earthen pyramids, aligned on electromagnetic gradients, where farmers brought seed before planting. In 1300 AD, the wealth and remarkable agriculture of Cahokia supported a population of about 30,000 — more than the city of London at that time. (Courtesy of Cahokia Mounds State Park Museum, Collinsville, Illinois)*

Plate 23. *Long before the pyramids, people erected over 10,000 standing stones in endless rows at Carnac, Brittany. The rows run alongside and parallel to an unusual geophysical frontier where seismic, gravitational, and magnetic gradients coincide. (Photo © Roslyn Strong)*

Plate 24. *Erected in a region of exhausted, infertile soil in southern England about 5,000 years ago, the 140-feet-tall Silbury Hill required fourteen million man-hours to build. Electric ground currents here are concentrated both at the peak and in causeways bridging the 'moat.' Over one week, electromagnetic measurements here fluctuated markedly. (Photo © Kaj Halberg)*

Plate 25. *The southwestern part of the enormous outer circle of standing stones at Avebury Henge, Wiltshire. Before the building of Stonehenge, people would come in the thousands to this place, and the owners of the henge became wealthy. (Photo © Kaj Halberg)*

Plate 26. *A partially restored stone avenue connects Overton Hill with Avebury Henge. The stones are like bar magnets, with all their north poles lined up in the same direction. Perhaps this was done to direct ions to the great stone circle at Avebury. Numerous large, glowing balls of possible plasma, electrified air, have been photographed around Avebury. (Photo © John Burke)*

Plate 27. *Mysterious ball of light at the end of the stone avenue, shown in Plate 26. The ball was not visible to the naked eye in this non-flash, daytime photo. Sophisticated analysis by top photo experts, using state-of-the-art image analysis equipment concluded that it was not a photographic effect. Rather, they say it was a real three-dimensional sphere about fifteen feet across, suspended above the ground and evenly illuminated from within – probably an aerosol. The same photographer has taken other similar photos, in the presence of the authors, on the opposite side of the stone circle of Avebury. (Photo © Nancy Talbott)*

Plate 28. *The bluestones of Stonehenge, transported 125 miles from Wales, were the most magnetic stones that the builders could locate. The 'C' shaped ditch, which truly defines a henge, would serve to concentrate electric ground currents that can be quite powerful at these carefully chosen, special sites. Stonehenge was in use for a thousand years and abandoned only after the discovery of fertilizer. (Photo © Kaj Halberg)*

Plate 29. *Djoser's pyramid was only the second structure in the world to be built of carefully cut and quarried blocks of stone, fitting tightly together. First a rectangle of limestone blocks was constructed, on top of that a slightly smaller rectangle, and so on for a total of six levels – a so-called step pyramid. (Photo © Kaj Halberg)*

Plate 30. *When the ancient Nile failed to flood, Egyptian fields were robbed of fertilizing mud, and famine soon followed. Fourth Dynasty pharaohs rushed to build the great pyramids at such sites and in such a way that they likely would have concentrated enough electric charge at the apex to ionize nitrogen in the air. This would provide free airborne fertilizer to what lay downwind – the farms. The photo shows the pyramids of Giza. (Photo © Kaj Halberg)*

CHAPTER 9

The Henges of Southern England

"Horses used to shy at them [the henge stones] in the dusk of evening and bolt down the bank."

—QUOTED IN *The New York Times*
"SURVIVORS OF AVEBURY" BY JASON GOODWIN

STONEHENGE IS A NAME THAT OVER TIME HAS BECOME ALMOST MAGIC, evoking words like antiquity, mystery, awe, power, and beauty. Such associations come to mind when you watch these massive, upright stones. Many of us are familiar with the annual news clips, showing thousands of modern Druids in white robes communing at Stonehenge each summer solstice. They claim a common law right to celebrate here the rites that their priesthood practiced for many centuries. This claim may well be true, but it is less well known that by the time the first Druids arrived in England, Stonehenge had been in use for thousands of years and then had already been abandoned for centuries. The Druids may have adopted it, but they had no more idea than we do today who built it or why.

For us, like many others, a long-running topic of conversation has been how such a magnificent work could have sprung up, seemingly from nowhere. In fact, Stonehenge is the pinnacle of a type of structure that had been evolving for centuries in Europe, in particular above the underground chalk aquifers of southern England. Once again, geology lay behind the design and location of one of the most famous types of giant megalithic structures: the henge.

Stonehenge really consists of two structures: a stone circle and a henge. A henge is simply a ring-shaped ditch. And for a thousand years before the stones went up at Stonehenge, people all over Europe were digging systems of ditches, called causewayed enclosures. We discovered that these structures also showed links with agriculture and energy.

INVASION OF FARMERS

Even the notoriously choppy waters of the English Channel were not enough to stop Neolithic sailors 6,500 years ago. Most likely using hide-covered boats propelled by paddles or oars, they crossed from Brittany and brought a new way of life with them. Their unfathomable effort of cutting down forests in order to scratch in the earth and their refusal to spend the day hunting must have looked like a sure recipe for suicide to the English natives. But, amazingly, they prospered. In fact, they were so successful that they created many problems for themselves. By about 3800 BC, whole sections of southern England were treeless. Heavy rains washed away huge quantities of topsoil from the hillsides. Continued farming of the same spot yielded less and less, and in as little as two or three years, the soil was drained of nutrients. The earliest farmers would just move to new ground. But this strategy could not last. The population of a peasant farming community grows at the rate of 0.5 to 1 percent per year. This means that within one thousand years a hundred initial farmers might number five million.[1] Also, the removal of the forest caused the ground to be waterlogged in low-lying areas, and blanket peat would spread. Once peat took over an area, it could not be farmed again.

Most arable land had already been farmed to exhaustion by 3200 BC, and England – like the rest of Europe – experienced a fertility crisis.[2] On Dartmoor, early farmers built stone rows, borrowing a page from their brethren in Brittany. These stone alignments were just two parallel lines, made of much smaller stones than the multi-ton behemoths at Carnac. Some are only four inches high. But this was a region with little stone available. As at Carnac, the size of the stones grew steadily near one end of the alignment. Some of the stone rows ran for miles and are still visible, for instance at Shovel Down.[3] In the Cornwall area, where stone was plentiful, dolmens were constructed – some strikingly similar to our New England rock chambers. Once again, a Neolithic population with exhausted soil turned to building megaliths.

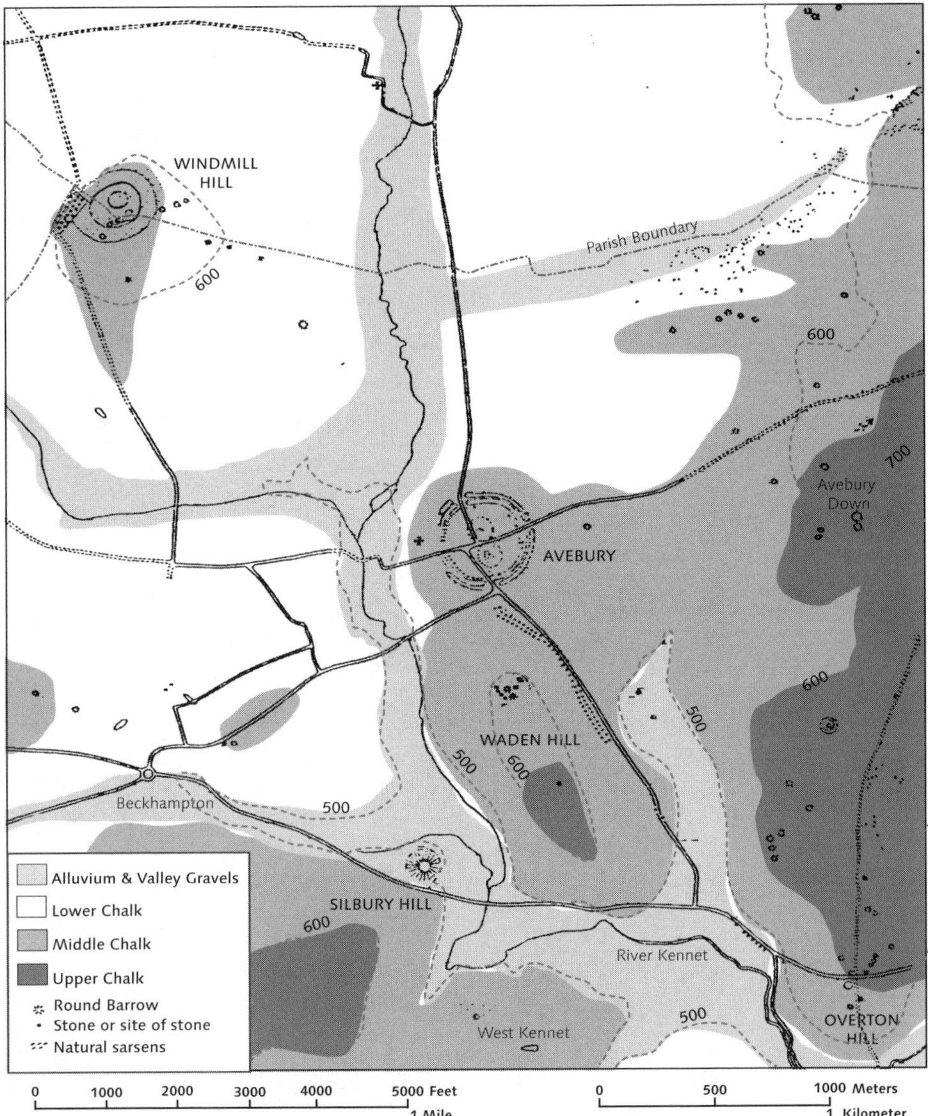

Figure 18. Geological map showing how Avebury, Silbury Hill, and Windmill Hill were all built on knobs of Middle Chalk protruding into the Lower Chalk layer. Here, the ditches of these structures are best able to cut the electrically conductive clay at the bottom of the Middle Chalk, thus concentrating telluric currents in the gaps (or causeways. The 'peninsula' and 'island' effects on these electrical discontinuities may help explain the unusually large magnetic fluctuations measured. (After I.F. Smith. *Windmill Hill and Avebury: Excavations by Alexander Keiller 1925-1939.* Clarendon Press, Oxford, 1965, p. xviii.)

A NEW TYPE OF MEGALITH

In one area, however, the megalithic structures took an unusual form. The Downs is a treeless grassland, differing from the rest of England by having numerous chalk hills. From 3500 BC until around 2000 BC, a range of megalithic structures was erected here, almost all concentrated on the chalk.[4] (Figure 18) They were not located near villages or even in convenient spots.[5]

On the Downs, where there are very few stones indeed, an entirely new model of Neolithic structure arose, named causewayed enclosures. They are areas, often with stone structures, surrounded by circular ditches hacked into the hard chalk, sometimes several in concentric

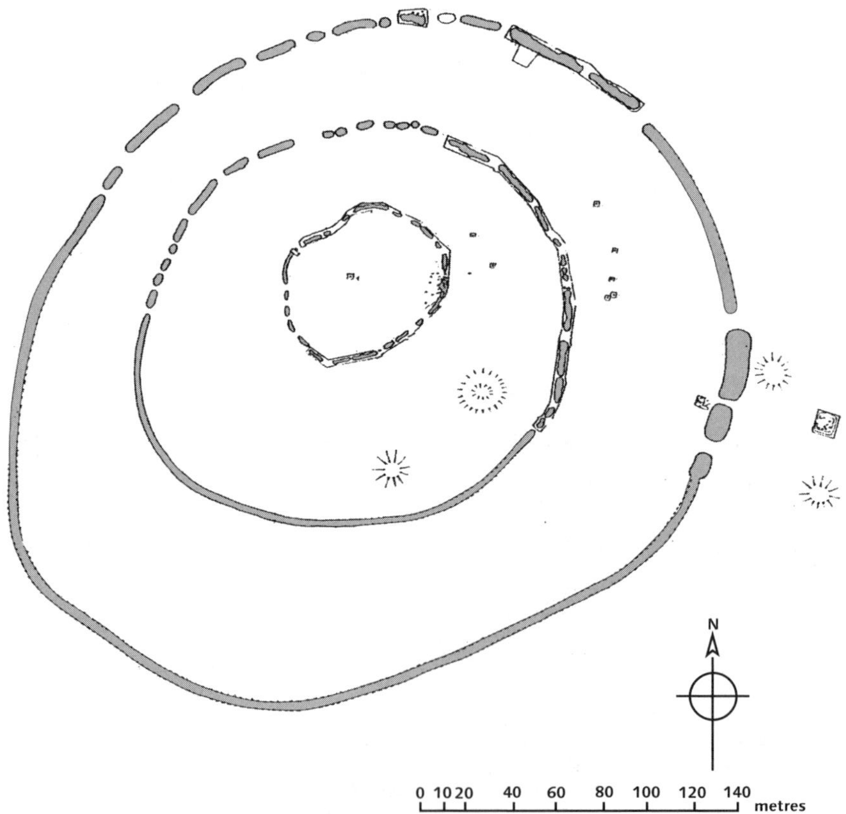

Figure 19. 5,000 years ago, a hundred thousand man-hours were spent digging concentric ditches at Windmill Hill. This out-of-the-way spot atop a hill was neither village nor corral. Excavations have found carefully cleaned seeds of emmer wheat, huge quantities of animal bones, and human skulls. (After I.F. Smith. *Windmill Hill and Avebury: Excavations* by Alexander Keiller 1925-1939. Clarendon Press, Oxford, 1965, p. xviii.)

circles. These ditches, called henges, were broken by one or several gaps, or causeways.

The largest of these early enclosures, on today's Windmill Hill in Wiltshire, consists of three concentric circles of shallow ditches, measuring over one thousand feet across (Figure 19). In the very center, atop the hill, the turf was removed to expose what geologists call 'the living rock' of chalk.[6] Since the enclosure was built, the entire hill has lost the top two feet of chalk dissolved by rainwater. As we shall later see, this process is crucial to the function of causewayed enclosures.

It is estimated that over 100,000 man-hours were invested in the construction of this enclosure, in an area with perhaps 4,000 people, including children. Today, most archaeologists state that enclosures were constructed to create sacred places where the shamans and perhaps the people would come for rituals. Certainly their theory may be correct, but on Windmill Hill, this space could have held 200,000 people – fifty times the estimated size of the local community.

And why did the builders almost always insist on hacking trenches into chalk rock, after dragging huge stones from afar? One academic theory is that these Herculean labors would foster a sense of community in the shared effort. Hard, unproductive work during a period of starvation is hardly the recipe for 'a sense of community.' Indeed, Neolithic England was not a Garden of Eden. People here were subjected to famine and bad health, such as missing teeth or no teeth at all, spinal arthritis, and malformed hips, arms, wrists, hands, or feet.[7] These people were struggling to survive, yet they devoted staggering amounts of time and energy to erecting structures.

Some archaeologists simply admit that they have no idea of the use or purpose of causewayed enclosures. Some declare that they were designed to keep cattle and other livestock penned in. Yet, there are so many breaks in the ditches at most enclosures that this explanation is hardly plausible. Many claim that the ditches were made for defense. Aside from the fact that there is no record of warfare during this era of the Neolithic, the large number of gaps in the ditches belies a defensive intent. Moreover, the banks made from the excavated soil from the ditches are on the wrong side – the outside – for defense. The ditches are usually only two to three feet deep, hardly enough to repel any invader. No residential houses seem to have been erected inside or even nearby. They were not located near villages or even in convenient spots. Therefore, historians conclude, their purpose was ceremonial.[8]

Certainly, ceremonial rites did take place at the enclosures. At Hambledon Hill, more than one hundred skulls with long bones were buried in just one part of a ditch.[9] Try to picture the spectacle of one or two hundred supplicants in solemn procession – perhaps at night, illuminated by torches – each carrying a skull and placing it in the newly excavated ditch. As at Carnac, skeletons and skulls were associated with regeneration and rebirth. At Windmill Hill, goddess figurines were buried in the ditches, making yet another symbolic statement to affirm regeneration and fertility.[10]

One of the first features completed was a series of small open air pits, lined with clay daub, along the innermost ditch. They look exactly like normal Neolithic grain storage pits, though shallower, ranging from 0.3 feet to a maximum of 1.4 feet deep.[11] Any seed placed in these pits would have far more contact with the atmosphere than in the average storage pit. Was the seed placed here, so that it could be exposed to electromagnetic forces?

At Hambledon Hill enclosure, traces of grain have been recovered, but it seems that crops were not grown here; "Grain arrived at the site already threshed and cleaned."[12] Pits of Windmill Hill enclosure contained pottery, so fresh when first used that it still carries imprints of the contents. The pots held emmer (a primitive type of wheat) and barley, the staples of that time. But there is something remarkable about the impressions. Sixty-one of them show emmer only – no impressions of any seeds of weeds or wild plants. The emmer was grown and harvested elsewhere, cleaned thoroughly, then brought to the site.[13, 14] It had not been ground for flour or other food but was brought in the form of seed. Some of the pottery was of a particular style, showing that it had been carried from a great distance.[15] It looks as if people were not just bringing food as offerings, but rather that they were carrying their seed there from a great distance.

INTERFLUVES AND GLOWING BALLS

The one factor that links all of the above into a coherent scheme is earth energy. The geology of the chalk downs is perfect for generating natural electrical current. It is all a matter of water moving through chalk. First, electrons are stripped off rain drops as they squeeze through the rock pores, in a physical process known as adsorption.[16] This means that the water molecules now have a net positive charge and have left a negative charge behind in the chalk. This effect is doubly reinforced when the water actually dissolves the chalk. A calcium carbonate molecule will split, causing a calcium atom with a double positive charge (an ion) to be washed away in the water, leaving behind a double negative charge in the rock. The total effect is chalk with substantial negative charge and water draining through it with substantial positive charge.[17, 18, 19] As we know, opposites are attracted, so this causes an electrical current in the ground. Just the movement of the electrically charged water alone will generate a magnetic field,[20] and we have measured both together at Silbury Hill after a night of heavy rain.

Combine all of the above with the fact that the seasonal vertical fluctuations of the water table in this chalk are some of the largest of any major aquifers in the world, up to a hundred feet.[21] Furthermore, the magnitude of magnetic field change is proportional to the porosity of the rock,[22] and chalk is extremely porous. Because of all these factors, the southern England chalk lands are some of the best terrain in the world for geologically induced electromagnetic fluctuations. Not surprisingly, henges were overwhelmingly built on chalk, just above a waterway.

Now we understand why the megalith builders in the Downs would first scrape the soil away until the 'living rock' of chalk was exposed directly to the air. The negatively charged rock would attract the positively charged atmospheric field lines, producing a scenario similar to what we found atop the Lost World Pyramid.

A final ingredient is our old friend, the conductivity discontinuity. Aquifers are not homogeneous affairs. They are made up of layers, separated by bands of clay. And in many places, including the Downs, geological uplift has folded these layers upwards until they reach the ground surface. Where the borders of two aquifer layers reach the surface at the same spot, we have an *interfluve*. Since aquifers have different abilities to conduct electrical current, an interfluve constitutes a conductivity discontinuity.

The giant Neolithic structures at Windmill Hill (c. 3200 BC), Silbury Hill (c. 3000-2700 BC), Avebury Henge (c. 2600-2400 BC),[23] Stonehenge (c. 2200 BC), as well as at Overton Hill, and numerous sites in Britain, were erected on knobs of Upper Chalk aquifers, intersecting with a layer of Lower Chalk aquifers; in other words, on interfluves. Famed archaeologist

Timothy Champion, in his classic text *Prehistoric Europe,* describes how for 1,000 years causewayed enclosures all over Europe were dug at interfluves.

In 1993, we measured powerful pre-dawn bursts of telluric currents on the highest ridge in Wiltshire, next to the Vale of Pewsey at Adam's Grave (a Neolithic mound), using our magnetometer and ground electrodes. Apart from the glorious views from here, this ridge is unusual electromagnetically. Just before summer solstice, we found the nighttime geomagnetic readings here weakened gradually by 400 gammas overnight – by far the largest nightly change we had ever measured. Then, within thirty minutes of sunrise, magnetometer readings came roaring back to normal. This activity is very unusual without a thunderstorm. A few days later, we measured a corresponding huge dawn surge in electric ground current. In latitudes like southern England, fluctuations of this magnitude are simply not supposed to occur, according to standard theory.

On this ridge, we first experienced one electromagnetic phenomenon, for which we have no explanation. Occasionally, mid-to-late afternoon will show an enormous jump in airborne electrical charge. We still do not understand the physical process behind this and, to our knowledge, no one does. But consider the following: One June afternoon atop the ridge that includes Milk Hill and rises right beside the oldest burial mound in the area, Adam's Grave, our assistant was learning how to use the electrostatic voltmeter. Suddenly, she cried out in some alarm, thinking she had done something wrong. She hadn't. As we all watched, some invisible power circled us and sent the needle of the meter climbing. We switched to the high setting, staring open-mouthed, as it went right off scale at 10 kilovolts per inch and stayed there. It never came down again. Back in the U.S., the manufacturer informed us that the unit was beyond repair and would be replaced.

Experts in ball lightning have estimated that DC electric fields of only 5 kilovolts per inch (half of what we measured) can be enough to produce a glowing ball of ionized air.[24] We saw no lights during the afternoon hours when our meter fried. However, from Milk Hill and Golden Ball Hill, parts of the same ridge, numerous observations of glowing balls of light have been observed, many of which were videotaped. One shows the entire top of Milk Hill, seeming to glow with a giant orange hemisphere.[25] At the foot of the hill, tiny balls of moving white light have been taped, during daytime. Another video shows a glowing blob at an altitude of several hundred feet, seemingly changing shape like a loose water balloon. Might similar lights have first drawn the early builders to these sites? Other magnetometer surveys, taken on different occasions on this hill, strongly suggest the presence of dynamic electromagnetic fields inside the ridge.

NEW PEOPLE – NEW IDEAS

Eventually, around 3000 BC, the quality of the soil around Windmill Hill deteriorated so badly that people abandoned the region. A few hundred years later, a band of farmers entered this area, at present-day Overton Hill, to have another try with the exhausted soil. Although rain was plentiful, the soil was too poor to support trees, and scrub and bracken grew all over the abandoned farms. Surely, to begin subsistence farming on such a spot would be an act of desperation. Conditions called for an extraordinary effort if they were to survive. They made this effort and what they did has baffled and intrigued the world ever since.[26]

One day, circa 2660 BC, perhaps just after the harvest of emmer wheat, a group of these new settlers gathered on a low-lying spot at the foot of a small chalk escarpment, where the Middle Chalk aquifer protrudes into a region dominated by the Lower Chalk aquifer. This means that the 'peninsula effect' of heightened electromagnetic fluctuations should be present here, as well as the usual interfluve effect. These farmers began removing the turf from the ground, cutting down until they had cleared a section of chalk bedrock, using a foundation of 'living rock.'[27] As at Windmill Hill, it seemed important that they expose this electrically charged bedrock.

The workers staked out a circle, 120 feet wide, and covered it with a thick layer of gravel from the adjacent River Kennet. Next, they piled cut sod within the circle. Atop this came four consecutive layers of soil, clay, chalk, and gravel, each layer about two feet thick. The final mound must have looked like a drum, 15 feet high by 120 feet wide. Around the mound, a 350-feet-long circular ditch was dug. This pattern is familiar from the causewayed enclosures, but this time it was on a far more impressive scale. The ditch was 40 feet wide and 20 feet deep, part of it dug through the adjacent chalk hillside. However, two sections of the hill were left intact, forming bridges over the ditch from the hillside to the central mound. As the long ditch was being excavated, chalk blocks from there were piled into a stepped network of walls comprising a honeycomb of interlocking cells, into which loose chalk rubble was dumped. As time has shown, this structure was marvelously stable.[28]

Year after year, the farmers hacked, dug, shoveled, and carried. They used sharpened deer antlers as picks to hack at the solid chalk. Shoulder blades of oxen were then used as shovels to scoop up the rubble and pour it into baskets. As the ditch sunk ever lower, these baskets probably had to be carried up ladders to ground level and then up again to the top of the rising mound. This kind of activity is consistent with the large numbers of broken bones among male skeletons from this time. The hill rose to an impressive fifty-two feet, the height of a five-story building. Then something remarkable happened. After the few thousand residents of the area had already invested an estimated one million man-hours of brute labor in this project, it was abruptly scrapped – just before being completed.

The ditch, still being excavated, was filled in. In Stage 2, a new, bigger, circular ditch was marked out, encompassing the previous one. Investigations have shown that no grass ever grew on the outer layers of the first stage. This means that once construction on Silbury Hill, the mound's current name, began, it would continue uninterrupted for an undetermined period – some estimate a century, others twenty years. It would all depend on how many workers were present and how much of the year they labored. Importantly, they never paused for so much as a single growing season.

Let us look at the human side of this situation. First, a group of farmers move into this valley, seemingly so desperate that they settle in an area that was long since abandoned as infertile. For some reason, they dedicate probably twenty-five percent of their manpower to building this hill, investing a million man-hours of hard labor. Then someone tells them that it is not big enough, that they will have to redouble their efforts. We propose that the people were willing to work on the mound because of some effect they wanted to achieve. So when, near the completion of the second stage, it became obvious that the desired effect still was absent, the people were willing to start all over again.

Stage 3 of Silbury Hill needed a completely new ditch. The old one was filled in (that must have hurt!) and a newer one circumscribed it, marking the perimeter of the final version of the hill – 380 feet. This ditch was more than twenty feet deep and seventy feet wide at the top. The new hill was a complete layering over of the old one, retaining its slope of thirty degrees. The top was left flat. Eventually, the work was completed. Silbury Hill stands undiminished today, almost 5,000 years later, attesting to the quality of the work. Holding a staggering 12.5 million cubic feet of chalk and soil, and covering 5.5 acres of ground, it rises sharply 130 feet above the low land (Figure 20).

Whatever the actual intention of the builders was, Silbury Hill became a dynamic structure (Plate 24). A dammed lake surrounding the

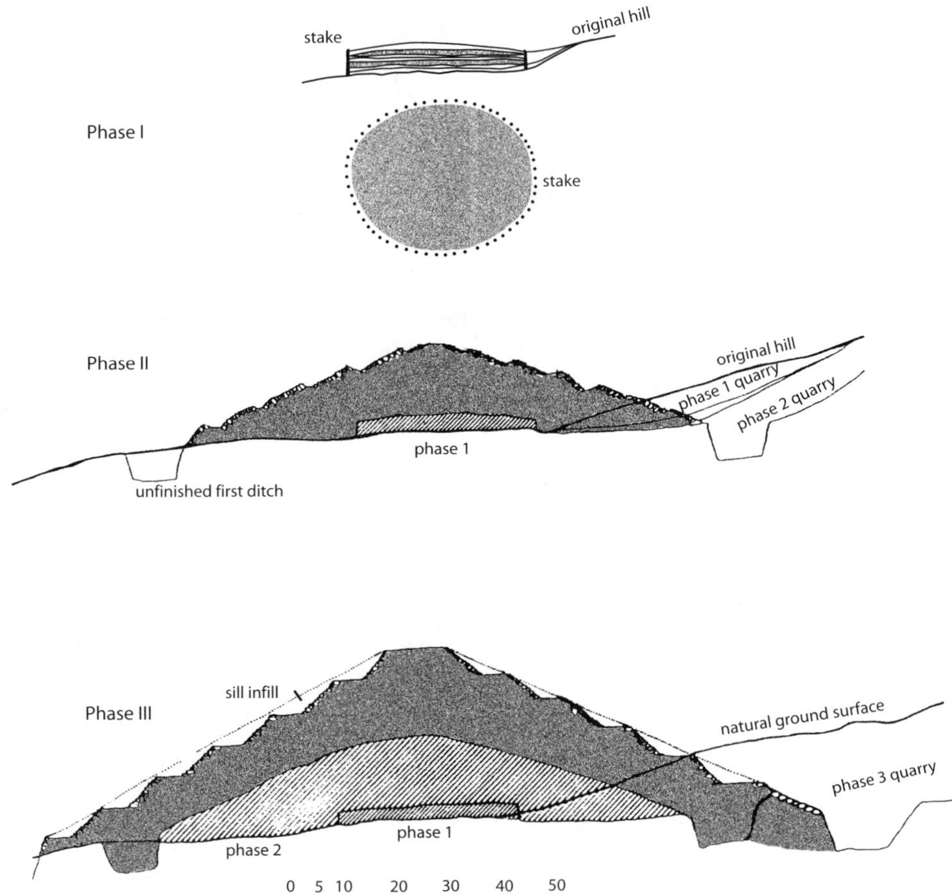

Figure 20. This illustration shows the different phases in the construction of Silbury Hill. The enormous labor that went into the building was invested by a population of a few thousand people, who were facing an agricultural crisis. Its design appears to concentrate and magnify natural electromagnetic fluctuations. After: Michael Dames. *The Silbury Treasure: The Great Goddess Rediscovered.* Thames and Hudson, London, 1976, p. 52.

base holds much less water each spring today than 5,000 years ago. Yet, the drainage of that water through the porous chalk as spring turns to summer still produces powerful electric charges that we have measured. On June 9, 1993, we planted geological electrodes at the top and bottom of Silbury. Keeping the top electrode in place at the center of the flat top, we moved the bottom electrode to different positions around the base of the hill, recording substantial ground charge. Table 3 shows our measurements.

This is how we stumbled into the secret of henges and causewayed enclosures. As described in Chapter Two, telluric currents are concentrated in the top few feet of the ground, and even a modest ditch will break their transmission, making them seek the path of least resistance. This is the importance of the causeway.

In this context, causeways are simply ground that the builders did not disturb. The original enclosures had many such gaps, but the ditch at Silbury Hill had only two. For telluric current, they were the paths of least resistance. And here before us, in the data from Table 3, was our proof: Causeways conduct electricity. 240 degrees west of the peak is one of its causeways, and our electrodes showed that ground current was pouring through it at more than double the rate of anywhere else.

Position of bottom electrode from summit	Reading in millivolts	Extrapolated to millivolts/km
Due North	-15.8 mV	-308 mV/km
Northeast	-21.9 mV	-289 mV/km
Due South	-11.2 mV	-148 mV/km
Southwest	-17.0 mV	-224 mV/km
240 degrees West	-47.8 mV	-630 mV/km

Table 3. Readings of electrical ground charge at different positions at Silbury Hill, England.

Was this why emmer wheat was placed in clay-lined pits near the causeways of Windmill Hill? Was this why the Native American mound builders also dug trenches, enclosing many acres, and placed mounds just inside the causeway entrance? In this way, the current would first be concentrated in the causeway, then again in the mound.

The electrical effects of Silbury Hill appear to be seasonal, linked to the movement of water through the aquifer. When we returned two months later, the readings were down substantially. Daily monitoring from July 31 to August 5, 1993, showed steady magnetometer readings and ground electrode readings from -72 mV/km to -110 mV/km, with one dramatic exception.

On August 2, it rained steadily all day, and water drained through the chalk surrounding the hill. The following day, our ground electrode readings had gone from negative 105 mV/km to positive 70 mV/km. This is a change of 175 mV/km overnight – very dramatic for geological processes. Even more exciting was the fact that these changes were accompanied by a corresponding alteration in the geomagnetic field. Magnetometer measurements atop the hill jumped first 150 gammas and then 300 gammas over the next few days. This shift is proportional to the change in telluric current, following a ratio well known to science.[29]

The most intriguing part about these measurements is not simply that they occurred in response to water draining through chalk, but that they were focused at the peak. Magnetometer readings at the base of the hill remained constant throughout this period. The top of the largest man-made mound in Europe, however, was turned into a concentrator of electromagnetic energy.

If achieving the effect we measured was the reason for building Silbury Hill, it would explain why, in this period of agricultural crisis, they prospered. Within a century or two, they would start an even more ambitious project – erecting the giant henge at Avebury.

THE GREATEST HENGE

Although Stonehenge gets all the publicity, Avebury is the largest and – many would say – the most impressive of all henges. As eighteenth century antiquarian William Stukely stated, "Avebury so doth exceed Stonhenge, as a cathedral doth a parish church." [30] Both sites, however, are masterful examples of the distinctive English contribution to megalithic architecture – the henge.

Avebury sits between Windmill Hill and Silbury Hill, perhaps half a mile from either. The design as well as its execution will awe any visitor. Inside the edge of a ditch, one thousand feet across and circular to within a few inches, stands a ring of rough, unfinished, standing stones (Plate 25). Originally, two hundred large stones were used for the circle but, unfortunately, most were destroyed in the eighteenth century. Only smaller ones remain, weighing seven to eight tons apiece. Some of the original stones weighed forty tons. Today, a small village sits inside the stone ring. In the ditch were four gaps or causeways, where the original bedrock was left intact.

No one knows how the builders surveyed the site to lay out the henge. The layout of Avebury is more perfectly circular than measuring with a rope would allow, because the stretching of the rope as it was pulled taut would create more error than is the case. Modern surveyors have difficulty being this precise. Without doubt, Avebury is one of the greatest engineering accomplishments of the ancient world.

The construction stretched over two or three centuries. It has been estimated that the work on the ditch and its bank alone required 1,560,000 man-hours, from a community of perhaps 4,000 people.[31] The ditch was as deep as thirty-three feet in some places, less in others. To excavate it, four million cubic feet of chalk, about 200,000 tons, were cut and carried up ladders. Clearly, there was something special about this place to the people who expended unimaginable efforts to create it. As with the causewayed enclosures, some historians maintain that Avebury was a defensive structure. If so, building the dirt wall outside the moat is poor planning – it gives your enemy the advantage of being able to shoot arrows down on you. Others theorize that the ditch was a cow pen.

The mound of material excavated from the chalk was simply dumped haphazardly in a hodgepodge of overlapping hills and ridges of various sizes; however, the ring of stones was so precisely circular that it defies understanding. This contradiction vanishes if one assumes that the precision work was for a utilitarian purpose, and that the piles of excavated dirt were debris from the ditch.

We do not dispute that ceremonies took place here. Similar to what was found at the causewayed enclosures, an abundance of human skulls and long bones were buried at Avebury Henge. In addition, many human mandibles, or lower jaws, were found,[32] just as at Carnac. These symbolic offerings have convinced many archaeologists that Avebury was a sacred ceremonial site. As we noted earlier, these offerings are typical fertility offerings, which would be in keeping with the henge as a generator of fertility.

From the central complex, two double rows of huge stones – today called avenues – were erected over two centuries, running out a mile or so. The West Kennet Avenue (Plate 26), still fairly intact, connected with Overton Hill – the site of a causewayed enclosure, stone circles, and a 150-gamma magnetic anomaly – to the southeast. The other avenue heads southwest to a spot near two huge, upright stones, called Adam and Eve. These stones are believed to be the only remaining ones from this avenue. The others were taken as building material over the centuries, and can be spotted here and there in the walls of houses inside the stone circle.

ELECTROMAGNETIC ENERGIES

Over several weeks, we conducted a massive magnetometer survey at Avebury, taking over 1,000 readings. On a hot August day in 1993, while bending down to take a reading, I suddenly heard an unidentifiable noise right behind me. Turning around, I saw a sheep happily munching my notes. A good hundred visitors to the henge that day were treated to the spectacle of a seemingly deranged man, shouting and sprinting after a harmless sheep – all the way across the circle. Luckily the records, though crumbled and smeared in saliva, were still readable.

During daylight hours in August, near the stones of Adam and Eve, we measured a mysterious surge in geomagnetic readings. Instead of the usual surge of about 20-40 gammas during one hour at noon, we found a 250-gamma surge, 150 gammas of which lasted several hours. We have never encountered a similar midday surge anywhere else, nor can we find mention of any in the scientific literature. But it makes us wonder if this was a spot known to the builders as rich in electromagnetic energy that they wanted to connect to the circle.

The daily magnetic readings all over Avebury die away overnight to a degree far greater than normal. Then, at sunrise, they suddenly come charging back. Like Silbury Hill and Windmill Hill, Avebury was placed so as to fill a knob of Middle Chalk that extended into the Lower Chalk formation (Figure 18). This placement will create the 'peninsula' effect, which heightens the fluctuations at this conductivity discontinuity.

At dawn, the ground current from the surrounding countryside is attracted by the henge, where the magnetic fluctuations are at maximum. It is well known that most of this current is transmitted near the ground surface. Digging even a shallow trench around the zone of magnetic changes creates an obstacle to the ground current. When electric ground current encounters the ditch, it will follow the path of least resistance, through the small opening of the causeway into the stone ring, now in a concentrated form.

Why did the builders of Avebury go to such trouble digging the henge up to thirty feet deep? Probably because underground electrical current travels best in wet layers of ground where the water in the aquifers of the chalk is stopped by a layer of clay. Throughout the 0.6-mile-long henge ring, the workers everywhere dug just deep enough to sever this electrically conductive layer and no deeper. This would force all ground current to concentrate at the causeways – where the stone avenues enter the central circle.

THE STONE MAGNETS

As mentioned, a double row of huge stones, called West Kennet Avenue, leads to the great circle of Avebury (Plate 26). These big slabs of sandstone, dragged from nearby Marlborough Downs, contain black magnetite, which make the stones magnetic. Retaining their original polarity from their formation deep underground, each stone acts like a weak but very large magnet. While the magnetism of the standing stones is not strong enough to noticeably deflect a compass needle, the more sensitive magnetometers show that the stones are indeed magnetic, as geological studies have confirmed. We recorded a particularly powerful jump in our magnetometer readings by holding the probe up to a fist-sized cluster of magnetite crystals, visible in one of the avenue sarsens.

If these stones were strictly for ceremonial purposes, the magnetic orientation of the stones would not be of consequence. However, the south pole of each stone faces the next stone in line as you move toward the circle. This arrangement means the north poles of the stones generally point south, which are opposing the geomagnetic field. Inside the main and the minor stone circles, the south poles of all stones point at the next stone in the circle, in a clockwise direction, with two exceptions. The stones at the two intact causeway entrances have their magnetic poles aligned with those of the avenue, rather than with the clockwise pattern of the circle, up to a ninety-degree difference from their companions in the ring. We measured all sixty-seven remaining stones, with an average of sixteen readings per stone. None had a detectable magnetic pole pointing in a direction that would contradict this pattern.[33]

Duplicating this arrangement in the lab, we obtained a result fully consistent with the laws of electromagnetism, but utterly shocking. The aligned magnets channel airborne ions in one direction! This is the same principle used at Chicago's Fermi Lab – America's greatest circular atomic accelerator. Today, physicists spend billions of dollars building circular tunnels with magnets, whose poles are aligned. We call them cyclotrons or colliders. They are built to move ions in one direction by making the magnets stronger and stronger as the ions move around the ring. Physicists use colliders to smash ions into targets so that they can study the debris of the collision and find the pieces that make up atoms. At Avebury, circa 2500 BC, the prehistoric engineers also seemed to know how to do direct ions, but doubtless in search of a different end result.

So our tests show that the avenue stones seem designed to channel airborne ions between them into the central circle. The orientation of magnetic poles of the stones in the ring would then contain the ions by circling them around and around within the ring, just like in modern supercolliders, but at a lower speed. Elsewhere in England and Scotland, and at Carnac, massive stone rows also end in stone circles. The rows usually connect the circles to water,[34] where air is easily ionized.

Just inside one of Avebury's major causeways was a small stone circle, surrounding a rectangle of standing stones – distinctive in these otherwise circular structures. Similar to what we saw at Windmill Hill, and later at Stonehenge, inside the rectangle were shallow, clay-lined pits. Neolithic farmers, we assume, put their seed into these pits, where the airborne ions would enhance them.

A potential problem would have been how to keep the ions from becoming 'stalled' at some point in the avenue by the opposite-acting poles of the magnetic stones. One way around this problem is to have the magnets become ever stronger. This solution may explain why the stones of the avenue grow in size as you approach the circle. This size patterning is also found at Carnac, Dorset, and other stone rows.

So is 'the collider' still at work today? Most of the stones are now missing, but consider the following and decide for yourself: One day in 1993, at the southern end of Kennet Avenue, Nancy Talbott, of Massachusetts, felt compelled to stop her car on a dangerous blind curve and shoot a photo of something she didn't consciously see at the time. When the film was developed, an anomalous ball appeared as a twelve-foot-wide sphere of glowing white, situated in the middle of the avenue (Plate 27). Analysis by photographic experts, using state-of-the-art computer enhancement equipment, concluded that it was not a photographic artifact but a real three-dimensional object, uniformly illuminated from within, probably an aerosol. In other words, it was much like a big ball of plasma, or electrically charged air.

Standing near the enclosure of Windmill Hill on another occasion, Talbott took two photographs of the opposite side of Avebury, revealing similar, if larger, glowing white spheres. None of her photos involved a flash. Glowing balls of light entering the ground or emerging from it, have also been seen several times at Avebury.[35]

In 1823, three stones were removed from Avebury's ring because "horses used to shy at them in the dusk of evening, and bolt down the bank." [36] Horses are not the only creatures to have been frightened by the stones. There are numerous cases of present-day visitors who get electric shocks from touching the standing stones.[37] We know of similar cases of shocks from New England rock chambers.

It is not hard to see why the people, who built this giant henge, would select this particular spot. Now let us look at what they got for all their effort.

CIRCULAR PROFITS?

Field experiments in Denmark have shown that the type of agriculture practiced by Neolithic people at this time, with the same crop grown on a field year after year, will exhaust the soil in a mere three years. Yields drop so dramatically as to not be worth planting. During the age of the megaliths, neither crop rotation, nor manure, nor any other fertilizer was known. One would expect such farmers to have abandoned the soil after a few years. But excavations have shown that, during this age, the average field was in fact farmed for ten years. Modern experts fail to understand how they got an extra seven years of useful production.[38] As was discussed in Chapter Six, if yields from traditional crops could be tripled just by exposing the seed to the energies at megalithic sites, then maybe we have found the answer.

In a pattern that must by now seem familiar, the people of Avebury quickly became rich. They were, in fact, richer than people anywhere else in Europe, with a single exception, the people of Carnac. Something was bringing wealth and distant visitors to the megalithic sites. Ceremonial stone axes from every Neolithic ax factory in England have been found in the Avebury area. In other words, people were streaming into this area from all over the country.

With the knowledge of Avebury's physics behind us, it is so much easier now to look back and understand the purpose of the long stone rows of Carnac. If indeed seed was being treated in the 'passage graves' there, the tiny, enclosed space inside these enclosures would severely limit production capacity. In air and on ground that could be electrified by magnetic and seismic forces, the stone rows could gather and conduct airborne ions into the stone circles at the end of the rows. If the rows acted like the ones at Avebury, they could have increased the seed treating capacity of Carnac to a level that would be sufficient to cover much of France.

At any event, in England as well as in France, an economic and social revolution seems to have swept aside the old ways. The egalitarian burials of earlier eras gave way to burials where males had dramatically more valuable grave goods than women and children.[39] Gold daggers of a striking craftsmanship so unique as to suggest a single master goldsmith have been found on the hips of males in both Avebury and Carnac. The men, probably the builders of the megalithic structures, were on the rise in the great game of social prestige.

Instead of dozens of long barrows with their egalitarian burials, we now have thousands of round barrows. Gone are the signs of any hereditary elite; instead we find the symptoms of a prosperous and populous new economic class. The round barrows were built, using the same clay layers as in Silbury Hill. Though thousands of them can be found on the Ordinance Survey maps of England, they occur overwhelmingly on the chalk aquifers. Most lie at the foot or on the crest of high bluffs, overlooking water. Modern aerial surveys of airborne electrostatic fields show that field strength peaks at such spots.[40]

Apparently, the men who built the huge 'generators of life' chose to be buried in a smaller version of these. Perhaps this was the Neolithic equivalent of embalming. The desire to be buried in a small Silbury Hill may not have been mere symbolism, but more like a modern individual opting for the expensive embalming and lead-lined casket.

A new era had dawned, reaching from Carnac to the chalk aquifers of southern England. Was a new elite created through enhancing seed and, thereby, producing more food for the masses? Is this how they and their people thrived on such terrible soil? The great authority Aubrey Burl [41] reports that in the centuries following 2850 BC, both population and agricultural output in the area increased. This growth would approximately coincide with the completion of Silbury Hill and the first stages of the construction of Avebury.

THE HENGE OF THE BLUE STONES

A few centuries of such prosperity would inevitably create envy. Thirty miles to the south, on Salisbury Plain, lies what is today called Stonehenge (Plate 28). It is situated on a peninsular knob of the Middle Chalk formation, where the chalk bedrock lies extremely close to the surface. Like Windmill Hill, this place was originally a causewayed enclosure. Its ditch was an irregular string of pits, steep-sided and flat-bottomed. The so-called Aubrey Holes are fifty-six uniform, more or less evenly spaced, clay-lined pits on the inner edge of the bank. But many more irregular pits were found littered all over the site. What was their purpose? It may help to look at the pattern of electrical ground current in the area.

In 1994, a survey of Stonehenge was made using a standard method of archaeological investigation, a ground resistivity survey.[42] The operator uses a device that sticks electrodes into the ground about one meter apart and then turns on an electric current. The device measures how well that artificially induced current travels across that one meter of open ground. The current's ability to travel is determined by how many free or easily displaced electrons are in the molecules of the soil. The more there are, the better the conductivity. Now, in our modern electric seed treatment, we have found that running a large voltage through the device moves out most of these free electrons, and it takes a while for them to be replaced through a connection to the ground. (If we have both a high- and low-voltage treatment to do the same morning, we try to do the low-voltage one first for this reason.) Something similar seems to be going on in Figure 21.

We know that natural telluric current travels primarily in the top few feet of the ground, and that a modest ditch will block much of it. Therefore, there will be less conductivity of telluric current across a ditch. Most of the current will flow through the unbroken ground of the causeway, as we measured at Silbury Hill. In Figure 21, we see a precise mirror image of this. The dawn surge of telluric current across the

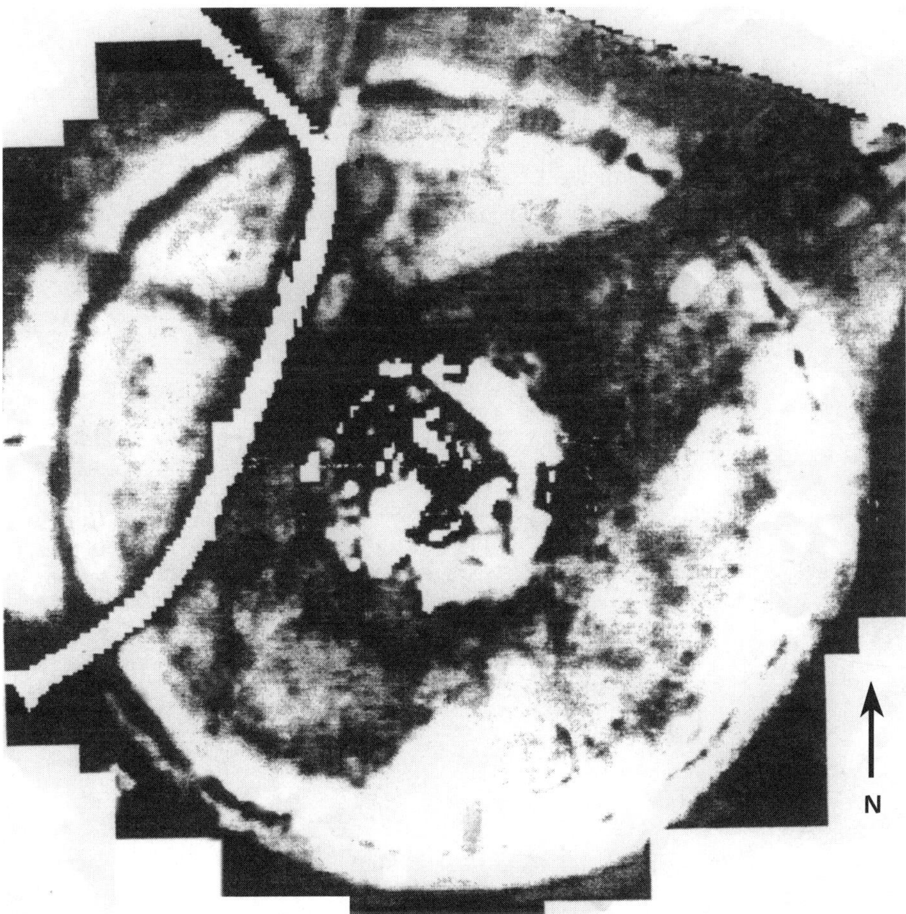

Figure 21. Electrical resistivity map of the ground at Stonehenge. Darker shading suggests earth that has had more natural electrical current run through it. Poor conductivity of the ring-shaped ditch forces the current to concentrate in the gap of the causeway entrance at the northeast. Inside, it fans out and hugs the stone circles (bright ring at center). The white, curving line to the left is the modern, paved visitors' path. (After R.M.J. Cleal, K.E. Walker, and R. Montague. *Stonehenge and Its Landscape*. English Heritage, London, 1995, p. 499)

causeway and into the henge's circle would knock out many free electrons, which might not be replaced at the time the artificial stimulus of the resistivity meter was introduced. This would cause the meter to code that ground as more resistive, or dark. Conversely, the ditch, which we know transmits natural telluric current worse than the causeway, would have less current move through it at dawn and retain more of its free electrons. This would cause the resistivity meter to measure that

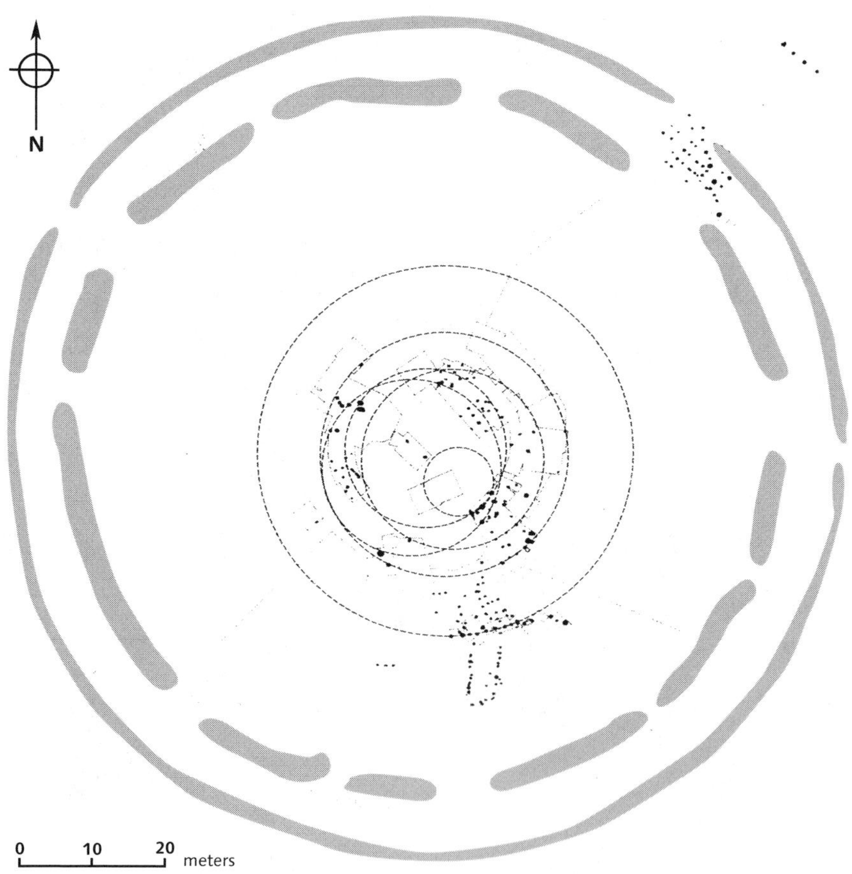

Figure 22. Illustration showing mysterious shallow pits at Stonehenge. Comparison with Figure 21 shows that these pits are concentrated in zones where the most ground current seems to have flowed, starting with the gap in the ditch, the open causeway in the northeast section where the most incoming current would have been concentrated. The lone exception to this trend is the rectangle of pits to the south, reminiscent of similar rectangles at Avebury Henge and Windmill Hill, which may have been used for temporary seed storage. (After R.M.J. Cleal, K.E. Walker, and R. Montague. *Stonehenge and Its Landscape.* English Heritage, London, 1995, p. 151)

ground as less resistive, even though we know it is the opposite; ditches block the flow of natural ground current. Viewed in this light, what Figure 21 shows is the expected result of telluric current rushing around the ditches, pouring in through the causeway (the gap in the ditch), and fanning out as it enters the henge circle.

Figure 22 shows a fascinating overlap with Figure 21. The small pits found in the ground at Stonehenge, which have never been explained, are located precisely where the rushing telluric current concentrates. If they once held seed, this is exactly where the seed would be exposed to the most telluric current, as it was concentrated by the structure of the henge. The only exception to this placement is the parallelogram of pits in Figure 22, which happens to be located directly above the largest magnetic anomaly at Stonehenge.

EXPANSION

After its construction around 3200 BC, the enclosure was in use for maybe 500 years. About the time that Avebury was built, the pits were filled in, and the site was abandoned. Was it no longer needed, now that Avebury was doing so well? Around 2200 BC, however, a group of people apparently decided that there was room for more than one horse at this trough. Burgess[43] states that Stonehenge was erected as competition to Avebury. Even the conservative authority Aubrey Burl considers it 'tempting' to view Stonehenge "as a monument intended to outdo Avebury."[44]

You would think that these new craftsmen would pick stones from Marlborough Downs, thirty miles to the north, littered with sarsens similar to those at Avebury. But they did not. They chose to go all the way to the Preseli Mountains in southwest Wales, 240 miles by land and sea. They traveled so far to collect bluestones, eighty-two of them, each weighing four tons.

English legend purports that Merlin the magician transported them through the air for King Arthur, for how else could they have gotten there? However, a quick glance at the map will reveal that the bulk of the route could be covered by raft along coastline and river, with the stones then pulled overland only six miles. As we saw in Chapter Five, far greater feats of transport were achieved in South America, and have also been duplicated by modern archaeologists.

What is special about this Preseli bluestone is that it is magnetic. The name comes from the blue color the stones acquire when wet. They consist of the mineral dolerite, a magnetic rock. In addition, the builders brought stones of softer volcanic and calcareous ash, as well as rhyolite – all minerals more magnetic than the sandstone sarsens at Avebury.[45]

At Stonehenge, eighty of the bluestones were placed in two concentric circles inside the ditch. Modern recreations, using only humans, poles, and ropes have determined how the stones could have been lifted into place.[46] But when the work was three quarters completed, the project was abruptly halted. Neither circle was ever completed. It may have become clear that they were not efficient enough.

If magnetism was indeed the characteristic sought out in henge stones, as was apparently the case at Avebury, then a basic miscalculation may have been made in using bluestones. While they have perhaps four times the magnetic field strength of the sandstone sarsens, the chosen bluestones were much smaller. Magnetic field strength is a reflection of how many magnetic field lines pierce one cubic centimeter of an object. But if the aim is to conduct airborne ions on a selected path, a small, powerful magnet may not outperform a weaker but larger one.

At Avebury, Stonehenge, and Carnac, the monoliths were not planted firmly in the ground. In the words of English authority Atkinson, "We know, however, that their builders were trying to achieve the maximum overall height with the material available, so that many of them stood in dangerously shallow holes and probably fell over at an early date."[47] This consequence had to have been expected by the builders, but was considered a justifiable risk to get as much of the stone above ground as possible. The higher a stone stands above ground, the more air and, therefore, the more ions it can interact with. So perhaps this was the crucial element in the effectiveness of a stone circle.

Did the stronger magnetism of the bluestones fail to compensate for their smaller size? After abandoning the bluestones, the builders went to Marlborough Downs in search of sarsen stones similar to Avebury's. They dragged them thirty miles overland to Stonehenge, dressed them into their now-familiar shapes, and erected a stone circle, surrounding a horseshoe of trilithons seven times bigger than the bluestones. What the latest stones lacked in magnetic field strength, they more than made up for in size. They may have had one quarter the field strength, but four to seven times the bulk. Unfortunately, the kind of polarity measurements that we performed at Avebury cannot be carried out at Stonehenge today due to the large steel revetments, inserted in the ground in the late 1950's to shore up the stones. The high magnetism in this steel would overwhelm any readings of the lower magnetism of the stones.

On the inside, the surface of the upright stones was smoothed, and even curved, at great expense. If the aim was to contain and swirl ions around inside the stone ring, the curving and smoothing of the inner surfaces made great sense. So did the later re-use of the bluestones for an additional circle at the center.

The causeway was widened an extra twenty-five feet, allowing more ground current to enter. Figure 21 may suggest why this was done. At the two o'clock position on the ring ditch, you can see that some resistance is still present where the chalk bedrock was originally cut. Yet the flow of dark over it shows that enough conductance was restored by filling in part of the ditch to continue the flow of ground current, perhaps eliminating what had become a bottleneck.

In Figure 21, the stone ring shows up as a white, nearly circular stripe at the center, indicating poor conductivity immediately around the stones. Unfortunately, here the steel revetments throw off all readings, so we may never feel confident about this zone. Otherwise, the dark, highly conductive region enters the causeway, fans out, and moves around the stone circle.

An 'avenue' of shallow, parallel ditches were carved out to connect Stonehenge with the River Avon, two miles away and quite a few feet below. Historians suggest that it was laid out as a processional route. Rather, the path may follow the course of cracks in the chalk where the ground charge would be at maximum strength. The water of the river could be expected to have positive electric charge compared to the negatively charged higher chalk that had drained water out. The tendency for positive ground current to flow from low water to high chalk would pull the beneficial negative airborne ions along with it. The ditches would stop the ground current from dissipating in all directions and channel it – and the ions – along the length of the avenue.[48]

RISE AND FALL

Before the giant stones were added to Stonehenge, people journeyed from all over England to Avebury. But now Stonehenge became a fierce competitor. Though they are only thirty miles apart, for the next few centuries people from the North went to Avebury and those from the South went to Stonehenge. People from the East and from the West chose either, half going to Avebury, the other half to Stonehenge,[49] traveling up to two hundred miles on foot.

Stonehenge ultimately won. Perhaps the architecture convinced people that it must be better. Maybe it gave better results because of constant remodeling of the design. Its engineers really never stopped tinkering with it. Perhaps the constant avalanching of the bank into the pit at Avebury reduced its effectiveness. Eventually people there gave up maintaining the ditch, and by 1900 BC, Avebury seems to have been abandoned.[50] Stonehenge then reigned supreme for the next several centuries. Round barrow burials around it include men laid to rest with jewel-encrusted daggers and other artifacts of wealth. So here, too, people grew rich.

Around 1500 BC, after at least seven hundred years of continuous use, people finally stopped coming to Stonehenge. Signs of wealth and aristocracy rapidly disappeared.[51] No invasion occurred at that time to have subjugated the builders and forced them to abandon their lucrative lifestyles. There are no signs of warfare breaking out. Why suddenly give up such a good life?

This question had nagged at us, regarding the sites we visited around the world. It was the Achilles Heel of our hypothesis. If these sites so improved agricultural production, why would they ever have been abandoned? What did happen at about the time that Stonehenge was abandoned was that fertilizer and crop rotation were introduced. English farmers realized that adding animal dung to their fields would keep them fertile.[52] This method seems to have been adopted almost simultaneously across the Old World – from China to Britain. At the same time, systematic growing of legumes like peas and Celtic beans was started, fixing nitrogen in the soil and regenerating its fertility.[53] With these discoveries, the farmer could now get improvement in yield without having to make the long trek to the henges. So, people stopped traveling to Stonehenge.

CHAPTER 10

Pulse of the Pyramid

"The river of Egypt is empty, men cross over the water on foot."

— NOFERTY (C. 1990 BC)[1]

SIR WILLIAM SIEMENS, THE BRITISH INVENTOR, WAS ONE OF THE GREAT nineteenth century pioneers of electricity. A basic unit of electrical conductance, the siemens (symbol: S), is named after him. Siemens related a literally electrifying experience that he had on the summit of Giza's Khufu Pyramid, The Great Pyramid, during the nineteenth century. One of Siemens' Arab guides called his attention to the fact that any time the guide raised his hand with fingers outstretched, they could hear an acute ringing noise. Raising just his index finger, Siemens felt an uncomfortable prickling sensation. As he tried to take a sip from his bottle of wine, he received a slight electric shock. Improvising on the spot, he moistened a newspaper and wrapped it around the bottle of wine. This converted the bottle to a Leyden jar, one of the earliest type accumulators of electric charge. The bottle became increasingly charged with electricity, simply by being held over his head.

When sparks began to emerge from the wine bottle, his Arab guides accused him of witchcraft. One of them tried to seize Siemens' companion, and in the ensuing struggle, Siemens touched the Arab with the bottle, giving him such a shock that he was knocked senseless to the ground, after which he scrambled to his feet and ran down the pyramid, shouting.[2]

As much as the giant pyramids of Egypt are like their megalithic cousins, they are also a special case. To understand why, we need to review the history of this unique kingdom. After the passage of the last Ice Age, approximately 11,000 years ago, the Sahara changed from desert to an enormous expanse of open grasslands, very much like today's African savanna further south. In these grasslands, big game species like elephant, giraffe, and antelope flourished, and so did the hunters who pursued them. Today, we find, scattered throughout the world's biggest desert, thousands of rock paintings from the millennia following the Ice Age. In graceful outlines, these paintings depict men with spears and bows bringing down antelope and gazelle.

About 4000 BC, however, the climate changed and the Sahara began to revert to desert once again.[3] Most of the game disappeared along with the water, and the hunters were forced to move in pursuit of what remained. But even at a desert waterhole, one cannot kill enough game to feed a band of people. The practice of agriculture had special appeal for humans in this situation. Farming had begun in Africa as early as 8000 BC, in the Ethiopian highlands.[4] But, as we have seen elsewhere, a special set of conditions had to exist for hunters to give up their nomadic ways and farm. In the Sahara, at this time, oasis agriculture started wherever possible. Nothing, however, could compare to the lure of the Nile.

THE GIFT OF THE NILE

For millions of years, the Nile River has been flowing 4,000 miles north, from the Equator to the Mediterranean. The almost mystical 'source of the Nile,' that reigned as the Holy Grail to an entire generation of European explorers in Africa, is in fact not a single source at all. Yes, Burton and Speke were right, Lake Victoria does empty into the White Nile, whose waters eventually flow past Cairo before fanning out to create a vast delta. But there is more to it than that.

From Lake Tana, in Ethiopia, originates another great river, the Blue Nile, winding its way through mountains to join the White Nile at Khartoum in Sudan. Along the way, countless other lakes contribute. From Khartoum, the combined Nile flows through high plateau and desert, rushing through a long series of great cataracts before entering what was Upper Egypt in ancient times. Flanked by massive sandstone cliffs on either side, the river passes through present-day Aswan before widening into a lush, green valley that stands in striking contrast to the empty desert sands surrounding it. About 500 miles long and, until recently, not more than two miles wide, this verdant strip would always have been the obvious homeland of choice for anyone in northeastern Africa who was looking for a place to farm.

In the days before the Aswan Dam, another well-known factor would give the Nile an almost supernatural allure for the farmer. Every year, the seasonal rains falling in the highlands of East Africa filter through a maze of streams and lakes week after week to slowly swell the Nile. By July, the river would flood Upper Egypt. Several weeks later, the surge would reach Cairo, inundating the valley there in October and November. This seasonal drowning of the land brought it back to life. When the waters eventually receded, a layer of water-soaked black mud was left on the fields. This mud consisted of decaying organic matter that was swept from the banks of countless tributaries. It was as fertile as soil can be.

Unlike most rivers that will deliver this agricultural bonanza at unpredictable intervals and levels, the Nile throughout the ages before it was dammed has generally risen with a regularity and predictability that made its flood plain the envy of farmers everywhere. This regularity enabled the inhabitants of the valley to dig irrigation ditches, raise houses, and level fields, confident that their work was not in vain. Thus, in contrast to the rest of the Northern Hemisphere, Egyptians planted in late fall and harvested in spring and early summer, waiting for the fall floods to again deliver their bounty of rich, moist soil.

So rich was this 'Gift of the Nile,' Herodotus tells us, that many early Egyptian farmers did not even have to plow the land. They merely cast their seed on the mud, then herded in flocks of sheep and goats, which trod in the seed. It was said that nowhere else in the ancient world did farmers reap such bountiful crops with so little work. Not surprisingly, soon after the Sahara began to dry up around 5000 BC, the valley filled with people eager to embrace this new way of life.

FAILING FLOODS

By 3100 BC, the competing kingdoms of Upper and Lower Egypt had been unified under a single king. The new capital was Memphis, located between the former kingdoms, south of present-day Cairo. Here a succession of kings, eventually known as pharaohs, ruled, and here they were buried along with their retainers and servants, entombed alive in the archaic rite of the God-King. Most of what we know about early Egypt comes from these tombs. In the early royal cemetery at Saqqara, they were called *mastabas*, rectangular, walled enclosures built of mud brick. Most scholarly histories of Egypt begin with the *mastabas*, move on to the pyramids, and generally follow the developments of architecture, war, and politics. In keeping with our theme, we intend to follow a chronology of the history of the land that was of far more importance to most Egyptians, be they peasant or noble.

About the time that the ditches were dug for the causewayed enclosure at Windmill Hill to combat the fertility crisis of southern England, a similar crisis struck in the Nile Valley. The wonderful predictability of the river floods failed. The climate of East Africa underwent a long-term alteration, causing the seasonal rains to decrease in volume.[5] From about 4800 to 3500 BC, the flood crest had remained fairly stable at about twenty feet above normal river level. Around 3300 BC, it dropped to ten feet, where it stayed for several centuries.[6] Ten feet was actually a better level for farming: just enough to inundate most of the valley, but not enough to sweep away villages or tear apart field systems. As this was the period when the nation had united under the pharaoh, everyone prospered. Royalty and their officers most likely enjoyed popularity and the general support of the people. All evidence suggests that during this early period in Egyptian history, the basic arts were developed. Formalized systems of government, writing, art, mathematics, and astronomy were all in place. There is no record of civil unrest during this extended period.

However, not long after 3000 BC, the Nile began to show occasional low flood levels of much less than ten feet, with little predictability.[7] We must bear in mind that unlike farmers in most other river valleys, the ancient Egyptians never saw the rains that fed their river. These rains fell hundreds of miles to the south in the mountains.[8] The farmers' decisions on how deep to dig this year's irrigation ditches or which fields to concentrate on, were based solely on expectations of flood levels similar to previous levels. Before and during the First Dynasty, c. 3050-2890 BC, the decisions had been correct. Then everything changed. Little is known about the exact reasons behind the ascension of the Second Dynasty, but is it only coincidence that at about the time the river fell, so too did the First Dynasty?

During the reign of the Second Dynasty (2890-2686 BC), the water level of the Nile continued to drop. By 2800 BC, its flood levels were down to five feet above normal, half the level of what had sustained the early unified nation. As Rushdi Said explains, "A single failure of the flood . . . could cause enormous misery and could leave an impact on the psyche of the nation."[9]

Shaduf, Egypt's ancient method of raising water in a skin bag tied to a levered pole, then pouring it into a field was as yet undiscovered. While some of the water loss could be offset by using Mesopotamian-style gravity-flow irrigation ditches, the loss of the fertile mud could not be replaced. The discovery of crop rotation and animal manure fertilizer lay far in the future and, after all, these farmers had never had any need of either.[10]

On other continents, the early farmers could move to another region when the situation worsened. That option was not possible in the Nile Valley. Walk one or two miles in any direction, and you would stand in rocky desert sands. There was no alternative to the valley. Doubtless, the situation did not deteriorate immediately. The soil was so rich that it probably could have been farmed for some years without soil replacement. Wheat and barley were the mainstays of their diet, and we have seen it take years to exhaust the soil in Europe and Mesopotamia, whose people were growing the same crops. Occasionally, the Nile would still flood to almost ten feet and renew most of the fields. Some unfarmed portions of the valley remained that were undoubtedly pressed into service. No one had ever seriously tried to farm in the water-rich delta to the north because swamps make poor farmland for wheat and barley, which both like good drainage. Pulses such as peas and lentils that could have restored nitrogen to the soil were traditionally grown separately under fruit trees in gardens on the high ground above the flood plain.[11]

After a few centuries of falling flood levels, soil exhaustion must have set in. No one knows exactly why the Second Dynasty fell, but at this point, it was replaced by the Third Dynasty (2686-2613 BC) under Pharaoh Djoser (or Zoser). During his reign, a seven-year famine swept Egypt.[12] Things went from bad to worse, as the waters of the river still fell. By now, it must have been clear to Djoser that the situation was desperate. Not only would his people continue to die if something was not done, his reign or the reign of his descendants would no doubt be rudely interrupted. His reaction? He built the first pyramid in the world.

Legend has it that Djoser summoned Imhotep to design a tomb for him. Imhotep was, seemingly, the greatest genius in the long history of Egyptian civilization. The Greeks credit him with creating the field of medicine. His actual titles included Chief of the Observers (head astronomer), Chief of Architects, Chief of Carpenters, and, most intriguingly for our purpose, High Priest of Annu.[13]

THE BENBEN STONE OF ANNU

According to the Egyptian creation myth, in the beginning all was water. Long before the Egyptians began worshipping the Sun God Ra, Atum was the Complete One, the creative power that lay behind the sun and everything else. He masturbated and from his seed sprang the pantheon of Egyptian gods and the first island, a hill called Annu. Gradually, the waters fell, and the island grew into all the land that we inhabit today.[14]

This is a natural myth from a people who lived on tiny 'islands' of raised land during the flood season, waiting for the waters to recede so that they could go out and seed the land. At the most sacred temple in Egypt, the Shrine of the Phoenix, stood a small hill revered as the original Annu

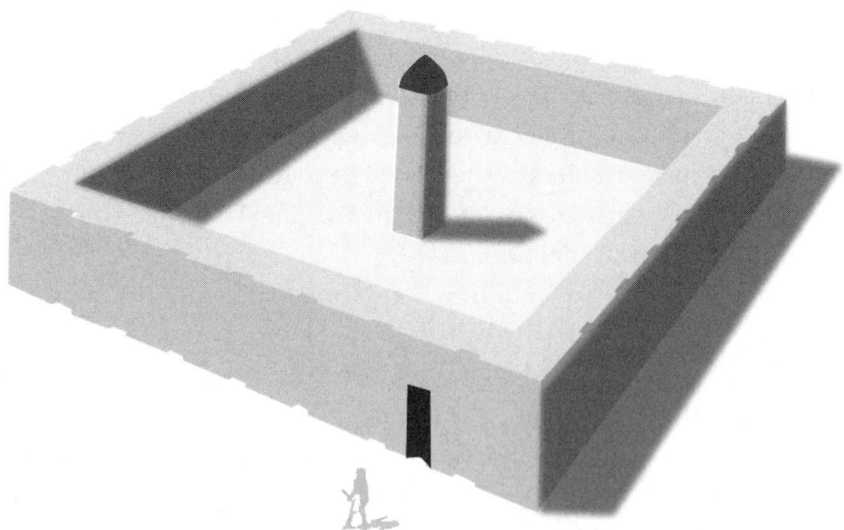

Figure 23. The holiest shrine of ancient Egypt, the Temple of the Phoenix, featured the Benben Stone – probably a conical iron meteorite – placed atop the Pillar of Atum, a limestone pillar on a sacred hill. (After Robert Bauval & Adrian Gilbert. *The Orion Mystery.* Crown Publishers, New York, 1994)

(Figure 23). The town here was later named Heliopolis by the Greeks in accordance with the sun worship prevalent at that time. On top of Annu was a stone pillar, connecting Earth to Atum. At the beginning of the pyramid age, an even more sacred object, the *Benben Stone,* was placed on top of the pillar.[15] Imhotep was head priest at this shrine, which was the center of Egypt's priesthood and its secret learning.

The Benben Stone was a black and pointed object, supposed to be the perch for the Phoenix, the mythological bird symbolizing Osiris, the god of rebirth[16] (Figure 24). The Egyptian name for the Phoenix was *bennu.* The root word *ben* was generally used to denote sexual, procreational, or seeding ideas such as semen, copulation, and, more important for us, fertilizer.[17] *Ben* is still used today in Semitic languages like Arabic and Hebrew to mean 'son.'

Figure 24. Phoenix, the Egyptian symbol of fertility is about to take flight from the Benben Stone. The traditional Egyptian Phoenix was represented by an ibis, a sacred bird. Originally, benben stones were probably pointed iron meteorites placed atop pyramids. Here, they could have ionized the electrically charged *khamsin* winds and spread fertilizing nitrates to farms downwind during a fertility crisis. (Illustration adapted from Papyrus of Anhai, British Museum, London)

The original Benben Stone is long gone. In later depictions, it is shaped like a pyramid, with the Phoenix perched on top. But in the earliest known depiction, its sides are bulging and rounded, suggestive of a conical shape rather than pyramidal. Many Egyptologists think that the Benben was actually an oriented iron meteorite.[18] Occasionally, an iron meteorite does not tumble on entering the atmosphere, possibly because an irregularity in its shape stabilizes it in one direction due to airflow over it. Whatever the cause, once the meteor becomes fixed in position, the heat of air friction ablates or erodes it in a non-uniform way. The result is a black, pear-shaped mass, similar to the giant Willamette Meteorite in the American Museum of Natural History in New York. Occasionally, the results of this process are more extreme and create a cone of iron, like the Morito Meteorite in the Institute of Metallurgy in Mexico City.

Since ancient times, meteorites have been widely regarded as sacred objects. For the Greeks, Delphi was the navel of the world. The *Omphalos* stone had been placed on the spot where Kronos originally had cast an object down, called *Zeus Baetylos,* which historians generally take to mean 'meteorite.' At Gythium, the inhabitants called their sacred stone *Zeus Kappotas,* meaning 'Zeus fallen down.' Pliny, the Roman historian, reports that "a stone which fell from the sun" was worshipped at Potideae and that others had fallen at Aigos-Potamus and Abdos. Black stones, said to have fallen from the sky, were revered throughout Syria. At Emessa (Homs), one of these meteorites was also conical. In ancient Phrygia, now in central Turkey, the goddess Cybele was represented by a stone like this, and her cult was carried by the Romans as far as France and England.[19]

The highest wish of all Muslims is to go on a *haj* (pilgrimage) to Mecca in Saudi Arabia because that city houses the central shrine, the Ka'aba. The Ka'aba in turn houses the most sacred object in Islam, a black meteorite, which was an object of ancient reverence even before Mohammed was born. When he and his followers took Mecca in 630 AD, he had the pagan idols surrounding the stone destroyed but embraced the stone itself as the sacred gift of God to Adam.

An Egyptian cult was, at least in Thebes, known to be centered on a black, conical iron meteorite. Therefore, we should not be surprised if one ended on top of the Pillar of Atum in the Shrine of the Phoenix.[20] Most important to our thesis, however, is what later happened to the Benben Stone.

THE ELECTRIFIED WIND

Unlike historians, who concentrate on the symbolic aspects of meteorites, we wish to consider the physical characteristics. All metals, including iron, are extremely efficient conductors of electricity. And, as we know from Silbury Hill and Central American pyramids, limestone can be a reasonably effective conductor if any trace metals are present. In the shrine of Atum where the pyramid architect Imhotep was high priest, it looks like an iron meteorite sat atop a limestone pillar. Let us look in detail, however, at the environmental facts in Egypt. Unlike the historical aspects of the country, these are still present today and can be studied and confirmed.

Every spring out of the Sahara comes a powerful wind, the *khamsin.* This wind is always hot and dry, and can blow at gale force for days at a time, often raising dangerous dust storms. Such winds are invariably filled with an excess of positively charged ions.[21] The similar *shirav,* the desert wind in Israel, is so laden with positive ions that it wreaks havoc with the mental and physical health of thirty percent of the population, who, incidentally, are effectively treated by inhaling large doses of negative ions.[22,23] Any time that sand particles are blown around by this type of wind, the friction between these semi-conducting silicon bits adds to the electrostatic charge of the wind. March through June is *khamsin* season in Egypt, with April and May being the peak months. The wind may strike without warning at any time during this period.

Earth is predominantly electrically negative, acting like a huge sink of electrons. As we also know, negative and positive ions attract and forcefully interact. Now consider the following quote from a scientist with long research experience in this field: "When a metal needle is subjected to a strong negative charge, electrons begin to escape rapidly from its sharp point; i.e., an avalanche of electrons of high kinetic energy is caused. This process is enhanced by molecules of atmospheric oxygen owing to the property of oxygen atoms to 'extract' electrons from metals. This phenomenon is known as electron or electrostatic emission. We used sharp needles because the quantity of electricity is directly proportional to the square root of the surface curvature."[24]

In other words, a powerful enough concentration of electric charge in the ground, if linked to a sharp metal object, will efficiently cause ionization of the air. This is essentially what happens when a church steeple draws lightning to it along the path of least resistance, a channel of ionized air. Actually, in the case of lightning, it is normally a positive charge in the ground that concentrates in tall, thin objects and attracts the negatively charged lightning bolt. But this process works even more efficiently in the opposite direction and helps explain why the occasional upward ground-to-cloud bolt is far more powerful.

Sometimes tall, thin objects will not attract a bolt of lightning. Instead they will begin to glow, and the glow can continue for hours. This is known as brush discharge or corona discharge and is essentially the same process as above but without the buildup and explosive release of charges that yield lightning bolts. Historically known as St. Elmo's fire, this glow has been reported countless times by sailors on mastheads and yardarms.

Now picture a pointed meteorite of nearly pure iron perched high atop a limestone pillar on top of a hill during a *khamsin* gale. The positively charged wind will tend to act like the base of the thundercloud, drawing negative charge from the ground. And, as we saw in Chapter Nine, water draining through carbonaceous aquifers like chalk or limestone creates negative electric charge in the ground. The *khamsin* blows during the months of the lowest Nile River levels. Thus, it occurs when much water has drained through the many-layered limestone aquifers that underlie the river valley and slope up its banks to the flanking high ground. There would be numerous electrons from the ground free to concentrate at the tip of the raised meteorite. The resulting build up of opposite charges would be a perfect lightning scenario in a thunderstorm, but Lower Egypt averages fewer than three thunderstorms per year. During the much more prevalent *khamsin* winds, however, the Pillar of Atum, topped by the Benben Stone, would present an ideal situation for visible brush discharge, presenting quite a spectacle to anyone looking at it.

To support our theory, let us look at the building of the Aswan Dam in the late 1960's, the last time Nile River levels crashed. For three years after completion of the dam, the water level of the Nile was extremely low, while the reservoir behind the dam filled up.[25] Water would drain down inside the limestone aquifers of the valley. Today, the Shrine of the Phoenix, with its Benben Stone atop a limestone pillar, no longer exists, but nearby the limestone walls of the Church of St. Mary point toward the sky. During these three years, in the *khamsin* month of May and during the following low-water months of the summer, the dome of the church would often glow and, occasionally, small balls of light would flow in the air. These phenomena were watched by thousands, sometimes for hours, and were often photographed.[26, 27, 28]

Research by J.S. Derr and M.A. Persinger revealed that the more persistent events here were "coronal type displays that were situated primarily over the apical structures of the church." [29] In other words, at the highest point near the site of the former temple of Annu brush discharges occurred where negative ground charge met positively charged atmosphere. Derr and Persinger were particularly struck by the "relatively sudden onset and persistence of the luminous phenomena." Their statistical analysis showed linkage with earthquakes hundreds of miles upstream, likely triggered by the immense weight of water building up behind the dam. Regarding the glow atop the church, they concluded that "the failure to find strong daily relationships between luminous phenomena and either seismic or geomagnetic activity suggests that some fundamental variable has not been accommodated . . . We suspect that this factor may be coupled to the mechanism that precipitates the actual occurrence of the strain field." We agree that the weight of the newly accumulated water in Lake Nasser caused the earthquakes, but we firmly believe that the persistent coronal discharges at the Church of St. Mary were set off by the falling water tables in the limestone along the banks, combined with the *khamsin* winds.

FERTILIZER FROM THE SKY

Imagine yourself for a moment as an Ancient Egyptian. When the Benben Stone, the most sacred object of your culture, would begin to mysteriously glow and give off sparks, it certainly fulfills one requirement of religion – the evocation of awe. Other effects of this electrostatic discharge for those standing nearby would be hair standing on end, and a tingling or prickling of the skin. Downwind from the pillar, many of the positive ions in the air would have been neutralized by the electrons escaping from the meteorite's tip. The sickening effects of the *khamsin* would be far less pronounced here. The smell of ozone could be detected and its presence would mean separation of the molecules in the air had taken place, a universal effect of electrostatic discharge. Our atmosphere is rich in nitrogen but in a form that plants cannot use. Brush discharge will ionize the nitrogen molecules of the air to become nitrates, readily usable by cereal plants.

In other words, these electrical forces would have another effect of vital importance to Egypt: They would provide airborne fertilizer. Farmers around the world have long recognized this effect. To the Hopi of southwestern United States, nothing is more welcomed than having their fields fertilized by the Sky God in the form of lightning. The Hopi know that crops will grow better when this occurs. This belief was dismissed by European-Americans as superstition. Now we know that lightning striking a small field can leave behind free nitrogen equivalent to one or two years' worth of fertilizer. Remember Tlaloc, the Aztec god of both lightning and fertility?

For much of the *khamsin* period, the Benben Stone might well have been releasing free nitrogen into the air. For much of the remaining year, it might instead have released negative oxygen ions, exactly like the ion generator in our living room. Negative ions themselves dramatically boost the growth, and probably the yield, of plants.[30] We do know that when a pointed structure is charged enough to glow, a plume of ions will be carried downwind.[31]

So may we not assume that a priest or a gardener in the gardens of the Shrine of the Phoenix would notice how much better plants fared on the downwind side of the temple? During most of the year, a gentle wind blows from the north. If this wind carried negative oxygen ions with it, the plants on the southern side of the shrine should have been taller, greener, and firmer. The *khamsin*, on the other hand, blows from the south and southwest. If it carried free nitrogen to the gardens or fields north and northeast of the temple, the plants here should have fared even better than those to the south during the period of low Nile floods. This fact would certainly have been linked with the glow at the tip of the Benben Stone during *khamsins*. Most likely, this glow would not occur throughout the *khamsin* but rather when ground and air charges were right. If these conditions happened only occasionally, it would likely be remembered that plants were affected when the Benben Stone glowed.

FATHER OF THE PYRAMID

Remember Imhotep, High Priest of the Shrine of the Phoenix, who was summoned by Pharaoh Djoser. Tradition holds that Djoser wanted a grand tomb for himself, but we think there was more to it than that. Since Imhotep was not only the Pharaoh's vizier but also Chief Architect and Chief Carpenter of the realm, he was the right man for the job.

All the previous pharaohs' tombs had been rectangular *mastabas*, entirely constructed of mud brick, and having flat roofs. Imhotep decided to build a vastly more expensive and labor-consuming structure. A rectangle of limestone blocks was erected, followed by a slightly smaller rectangle of stone atop the first one, and so on for a total of six levels, a so-called step pyramid. This pyramid, like all Egyptian pyramids, was not built by slaves. Most historians today agree that they were erected by volunteer farmers during the flood months of October and November when they were idled by the inundation of the fields. Somehow during the decades of the great famine, these farmers were motivated to spend several months year after year, building the biggest structure ever made by man until then.

Djoser's step pyramid (Plate 29) was the first structure in the world to be built of carefully cut and quarried blocks of stone, fitting tightly together. The limestone used was cut locally on the West Bank, where much *dolomite* is found.[32,33] Dolomite is limestone with a manganese content above twenty-five percent. Our own laboratory tests have shown that dolomite is a good conductor of electricity. We have found that the conductivity of limestone is in direct proportion to its content of manganese. So Djoser's pyramid was made of highly conductive blocks, fitted tightly together.

We will never know if anything lay atop this pyramid, but we do know that the later, pointed pyramids had, at their apex, a stone of polished black granite, called benben, bearing the image of the Phoenix. Because of their pyramid shape, they were named pyramidions by Egyptologists. If the pyramid of Djoser had merely a symbolic granite image of the original Benben Stone, it might have functioned much like Silbury Hill. If the benben was real iron, it would produce a greatly magnified version of the ion effects that we suppose were present at the Shrine of the Phoenix, possibly enough to substantially improve crop yields. Imhotep probably knew of this effect.

Another symptom suggesting that the intention of building Djoser's pyramid was to have a generator of ions and free nitrogen, is the fact that it was ninety percent completed when the builders started all over again. As at Silbury Hill, someone decided it now had to be twice the size. Was some effect absent that Imhotep had hoped to achieve? The grandeur of its appearance could have been judged long before it was ninety percent complete. But its physical effects could only be judged once it was essentially finished.

TOMBS WITH NO BODIES

At this point you may ask yourself why we are going to such extent to analyze the electrical functions of structures that almost all archaeologists agree were giant tombs for the pharaohs. We are informed that the pyramids were built on the west bank of the Nile because it is the land of the setting sun and, therefore, of the dead. This theory certainly sounds plausible.

But, to begin with, there is nothing preventing the giant pyramids from having had two functions, one symbolic and one practical. Now, one essential problem with the theory that these huge structures were built specifically to be royal burial places is there are no bodies to support it. No pharaoh's remains have ever been found in the large pyramids. Not even one. The only remains found proved to be later insertions of corpses, hundreds of years later, similar to the pattern we have seen in France and England.

Most of the so-called 'burial chambers' were built around a giant sarcophagus and afterwards, they seem to have been sealed up forever. Some of the chambers were even left unfinished, mere caverns of crudely chipped limestone. Some of the finished ones were sloppy in execution with measurements and angles showing tremendous variation – nothing like the incredible precision, which typified the pyramid itself. At most, in these chambers, archaeologists found an empty sarcophagus often without a lid. 'Grave robbers,' we are told, carried off the rest. If the burial chambers were optional, and not the primary purpose for construction, that would explain why some of the chambers would be left unfinished or unused.

Sekhemkhet, Djoser's successor, began building a pyramid that rose only twenty-five feet before being abandoned. When it was discovered in 1951, the chamber beneath it created a great stir because it was discovered intact. A corridor leading to the chamber was filled with thousands of animal bones, reminiscent of Avebury. When Dr. Zakaria Goneim broke through the intact walls of the chamber, he found an underground hall, hewn roughly from the limestone bedrock. In the center there stood an exquisitely polished sarcophagus of alabaster, completely sealed. An ancient bouquet, left atop it, had crumbled to dust – proving that the sarcophagus had remained undisturbed. Four months later, an official delegation of government representatives, archaeologists, and journalists were gathered for the formal opening. As they collectively held their breath, the alabaster lid was levered off the sarcophagus with much trouble, and everyone stared in shock. This never-opened sarcophagus was completely empty. Clearly, the 'burial' was symbolic.

As we have seen over and over again, common themes, symbols, and structures arose repeatedly in different locales. The 'burial chambers' inside or below the Egyptian pyramids resemble the Meso-American model where pyramids were erected over sacred caves, quite likely fertility caves. And we have seen repeatedly in France, England, and Meso-America the use of corpses or bones for internment in a mound, rock chamber, or pyramid. We have also seen the close connection between the bones/corpses and fertility. Might the Egyptian kings have helped fire up their people's enthusiasm for building the pyramids by promising to lend their own corpses to increase the power of the pyramid? The kings would likely have been informed by the engineers that the actual presence of the corpse was unnecessary. Perhaps though the elite preferred to let the common people believe this useful myth.

The theory that these early, giant pyramids were tombs arose from our knowledge of the much smaller pyramids of the Fifth Dynasty (2494-2345 BC), built several hundred years later. Extensive numbers of hieroglyphics inside these later and much smaller pyramids, as well as a temple of that era, inform us that the later, smaller pyramids were in fact the final resting place of their pharaohs, and some of them did contain bodies.

If indeed the great pyramids of the Fourth Dynasty (2613-2494 BC) were burial chambers, then why are they mute regarding the pharaohs whose monumental tombs they supposedly were? No one has found the names of the pharaohs inscribed on these structures. Our knowledge of who constructed a particular pyramid has come from workmen's scribbles in obscure places inside the pyramids, or from later Egyptian records, unlike the Fifth Dynasty pyramids. It is not that the Fourth Dynasty was shy about public use of hieroglyphics. Pillars, called *stelae,* were erected throughout Egypt, covered with hieroglyphics that in detail describe various historical events and the accomplishments of the reigning pharaoh.

The situation is reminiscent of the Mayan pyramids described in Chapter Four, which became non-utilitarian, political structures after a century of foreign occupation where the society dissolved and the knowledge of the fertility generating properties of the pyramids seemingly was lost. In Egypt, the great pyramids were hurried into service from circa 2650 BC to perhaps 2400 BC, while the Nile continued its poor performance, only to reach a terrible low around 2200 BC. For nearly two centuries, it failed to flood at all. Devastating famines finished the pharaoh's mandate, and the Egyptian people turned their back on the Fourth Dynasty. From the ensuing chaos, the Fifth Dynasty eventually arose, two hundred years later.

The pyramids of the Fifth Dynasty are truly puny, compared with the earlier giant structures. They were of the shoddiest construction and haphazardly scattered. Instead of solid, quarried blocks, these later imitations had a core of loose piles of rubble that have shifted over the centuries, causing many to collapse. There is no reason to link these tiny affairs with the earlier pyramids.

Egyptologist John Wilson of Chicago University summed up his view of this degeneration as follows: "The several pyramids of the Third and Fourth Dynasties far surpass later pyramids in technical craftsmanship. Viewed as the supreme efforts of the state, they show that the earliest historical Egypt was once capable of scrupulous intellectual honesty. For a short time she was activated by what we call the 'scientific spirit,' experimental and conscientious. After she had thus discovered her powers and the forms which suited her, the spirit was limited to conservative repetition, subject to change only within known and tested forms." [34]

PYRAMIDS OF SNEFERU

Historical records do not give any reasons for the fall of the Third Dynasty, but we do know that it suffered famine. The Fourth Dynasty came to power with a mandate that they were determined to carry out at all costs. Huge irrigation projects were undertaken, dams were erected to hold water for the off season, and irrigation systems were improved to maximize what water was available from Nile floods and reservoirs. Such intervention was needed because the Nile was still low. As previously mentioned, until now the Nile delta had been ignored as arable land. But during the Fourth Dynasty, vast royal estates were cultivated here to augment the nation's agricultural production in order to provide food for the laborers building the pyramids.[35]

The Fourth Dynasty was the zenith of the Pyramid Age. In terms of volume, eighty percent of all Egyptian pyramids were erected during the 120 years of this dynasty. The first king Sneferu ('Bringer of Beauty') was credited with not one but three pyramids, none of which, incidentally, held his remains. And again, Sneferu did not put his name on any of these three structures. He expanded and completed the step pyramid at Meidum, converting it to the first example of the 'true' pyramid structure with four flat, smoothly sloping sides rising to a common point.

He also erected two pyramids just south of Saqqara, at Dashour, by far the biggest built until then. The larger of the two, the Red Pyramid, is made of a reddish limestone, loaded with iron, and therefore, highly conductive to electrical current. Any *khamsin*-driven movements of ions or airborne nitrogen from the peaks of this pair would likely have settled around the capital of Memphis, reinforcing the effect of Djoser's pyramid on the crops of the capital area.

The pyramid builders were, of course, fully human and prone to less than perfect ways. It appears that the outer casing of the Meidum pyramid collapsed when it was about halfway finished. In the famous Bent Pyramid, the builders altered the angle of ascension midway through construction. Some Egyptologists think that the builders, after the collapse at Meidum, feared that the steeper angle could not support the building. Others hypothesize that the pyramid had to be completed in a rush, and the lower angle would require less stone. 'Rush' definitely became the word most identified with construction of the Fourth Dynasty pyramids.

Building three giant pyramids during a reign of about thirty years had to tax Sneferu's people sorely. The two at Dashour alone contain 7.6 million tons of limestone blocks. With an average weight of 2.5 tons each, this meant 3,000,000 blocks. If the construction went on every single day during his thirty-year reign, it would amount to the cutting, moving, and setting of 300 blocks a day. But we must remember that most of the work probably took place during the two or three months a year when the farmers were idle. This would equal approximately 1,500 blocks a day, assuming the construction took place over the full thirty years.

The common image of slaves hauling stone sledges, driven by the lash and enduring horrible conditions was probably reinforced by these kinds of computations. Who else would put up with this? And though we know that the pharaoh was an absolute ruler, Sneferu had just been swept into power to replace the previous 'absolute' ruler. He was well aware of the limits of such power during food shortages. To us, the most amazing thing about Sneferu is not that he was able to perform these engineering miracles but rather that historical records indicate he was well-loved by the very people whom he seems to have driven remorselessly.[36]

Said tells us, "In case of good governments which stored grain surpluses in years of good Nile to dispense them in years of poor Nile, conditions became worse only when there were two or more consecutive failures of the Nile."[37] Of the six recordings of Nile flood levels from the reign of Sneferu, half of these were below the proper level.[38] If this sample is representative of the period, then Egyptian agriculture, and, therefore, also its people, were in dire straits indeed. Suddenly it becomes much easier to imagine how the people could be mobilized for these Herculean labors, suggesting that they may have known of some specific benefit from them.

THE GIANTS OF GIZA

Sneferu's son was Khufu, called Cheops by the Greeks. He built what remained the largest structure in the world for 3,000 years until the Great Wall of China was constructed. The Khufu Pyramid, called The Great Pyramid (Plate 30), stood 500 feet high, the height of a fifty-story building. Its perimeter is more than a thousand yards, and its 570,000 square feet base is enough to hold the Cathedrals of Florence, Milan, St. Peter's, St. Paul's, and Westminster Abbey put together.[39]

Khufu's successor, Khafre, built a near twin. Just a few feet shorter, but erected on a slightly higher part of the plateau, it looks even taller than Khufu's. As a monument to himself, Khafre would have produced a far more imposing appearance by standing this giant off on its own, but he chose not to. Was there a functional reason?

The third pyramid of Giza is associated with Khafre's successor, Menkaure, and is truly puny compared with the other two. At 200 feet, it is less than half the height. At 0.6 million tons, it is one tenth of the mass of the others. Giza is a large limestone plateau, artificially leveled to support not only Khufu's champion but also the other two pyramids. Earth moving on such a gargantuan scale, preparing a level base for all three pyramids, suggests a master plan, with Imhotep, traditionally suggested as the author, by now deceased. John Legon convincingly shows that the three were planned as a group.[40]

ELECTRICAL PYRAMIDS

Did Imhotep design the pyramids to act as giant ion generators for most of the year and a nitrogen producer during the *khamsin* months when the Nile was at its lowest and, therefore, all the farmland was exposed? The pyramids were, in fact, constructed in such a manner as to accomplish just that.

All pyramids were built with an electrically conducting core of coarse local limestone with a high content of magnesium. These core blocks are what we see today, the sides being anything but smooth. This is because, in the thirteenth century Moslems tore off the outer casing and used it for building materials as the start of an ultimately abandoned effort to destroy these pagan structures.

In only a few spots can we still see the beautiful, snow white, fine Tura limestone that made up the outer casing. This stone was finer-grained than the stone employed in the core, and polished to a sheen. It came from the East Bank of the Nile and had to be ferried across. Most of the giant pyramids were ultimately encased in Tura limestone. We find this intriguing because, in contrast to the core limestone, Tura limestone contains only traces of magnesium and therefore is an extremely poor conductor of electricity.[41] In fact, it acts as an *insulator.*

What we have then, is a massive pyramid of electrically conducting limestone blocks, cloaked in an insulating outer layer. This combination would prevent any of the electrical charge in the core from leaking off into the air. The Tura casing blocks are one of the greatest engineering wonders of the ancient world. They were polished on all sides to within 1/100th of an inch accuracy, fitting together so snugly that you cannot get a razor blade between them, five thousand years later. The labor involved in this creation was extraordinary. The builders either had an incredible fetish for precision, or this precision was functionally important. With this tight insulating cover, the only place the electrical charge could leak out would be through the benben, at the apex of the pyramid. All negative charge, spread throughout its entire base would be concentrated in one pointed capstone, possibly of pure iron.

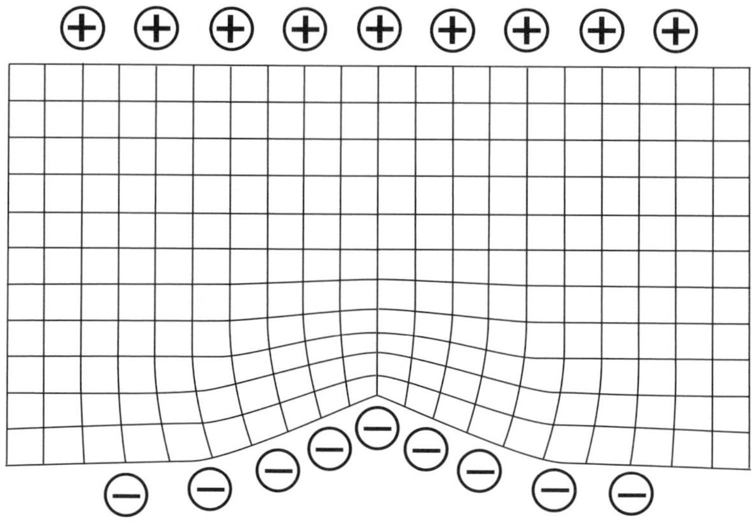

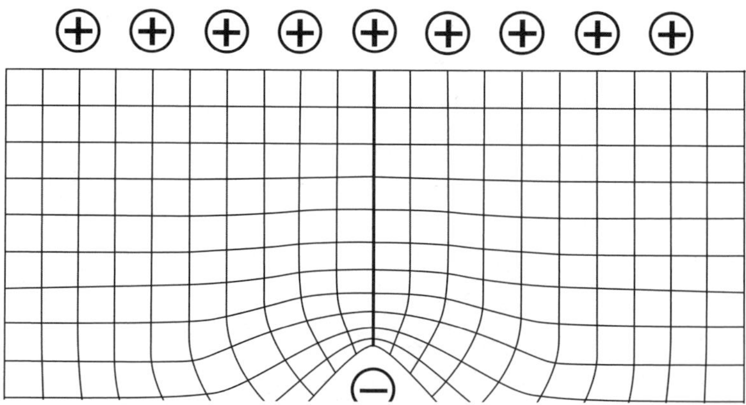

Figure 25. Any protrusion on the earth's surface causes bunching of field lines of the atmosphere's positive electric field at the peak, which means a concentration of positive electric charge. The sharper the angle at the peak, the stronger the concentration. This may explain why Egyptian builders erected pyramids with the steepest angle they could maintain. Negative electric ground current flows in through the base. The larger the base, the more negatively charged electricity is present during the season of the *khamsin* winds. (Illustration from R. Reiter. *Relationships Between Atmospheric Electric Phenomena and Simultaneous Meteorological Conditions. I.* Air Research and Development Command, U.S. Air Force, 1960.)

On a pointed hill or mountain, the positive electrical field lines of the atmosphere that would normally be spread out across the entire base are concentrated at the peak. Scientists have also found that in Nature by far the highest electrical readings were obtained near the edges of sharp rocks.[42] We have confirmed this effect on The Lost World Pyramid in Guatemala.

Figure 25 shows that the steeper the peak, the more concentrated the electrical field lines. In this context, it is fascinating to note that Egyptian pyramid builders always tried to make the steepest slopes that would remain structurally sound, namely about 52 degrees.

Today, nearly all the insulating cover of Tura limestone is gone. The exposed limestone blocks of the conducting core should dissipate much of the electrical charge from inside the pyramid through their exposed edges. After all, any sharp edge concentrates electrical charge and becomes a point where the charge bleeds off into the surrounding air. Again, this has been confirmed by us in Guatemala. So, in fact, any electric charge moving up inside the pyramid today, will never reach the peak in any large concentration. Additionally, the pointed capstones have long since disappeared. At the summit of the Khufu Pyramid we now see a flat area, several meters across. This current, deteriorated form cannot concentrate charge to anywhere near the degree of the original pointed top. But even in this ruined condition, electrical effects have been noted, for example the story at the beginning of the chapter about Siemens's electrically charged bottle.

AIMING THE IONS

The geographical siting of the great pyramids was hardly a coincidence. All were built atop a limestone plateau on the west side of the valley. They all had contact with the limestone bedrock. Underground, the rock strata tilts downward toward the river, forming the edge of the Nile Valley aquifer.[43] As we know, water moving up and down in aquifers creates electrical currents.

A question that has long puzzled Egyptologists is why Giza was chosen as a building site. All previous pyramids were built well to the south, near the capital of Memphis. Why then this change? Once again, a study of the topography involved may shed some light.

The Giza plateau itself is the intersection of two major limestone layers: the Mokkatam and the Maadi.[44] All three pyramids were placed in

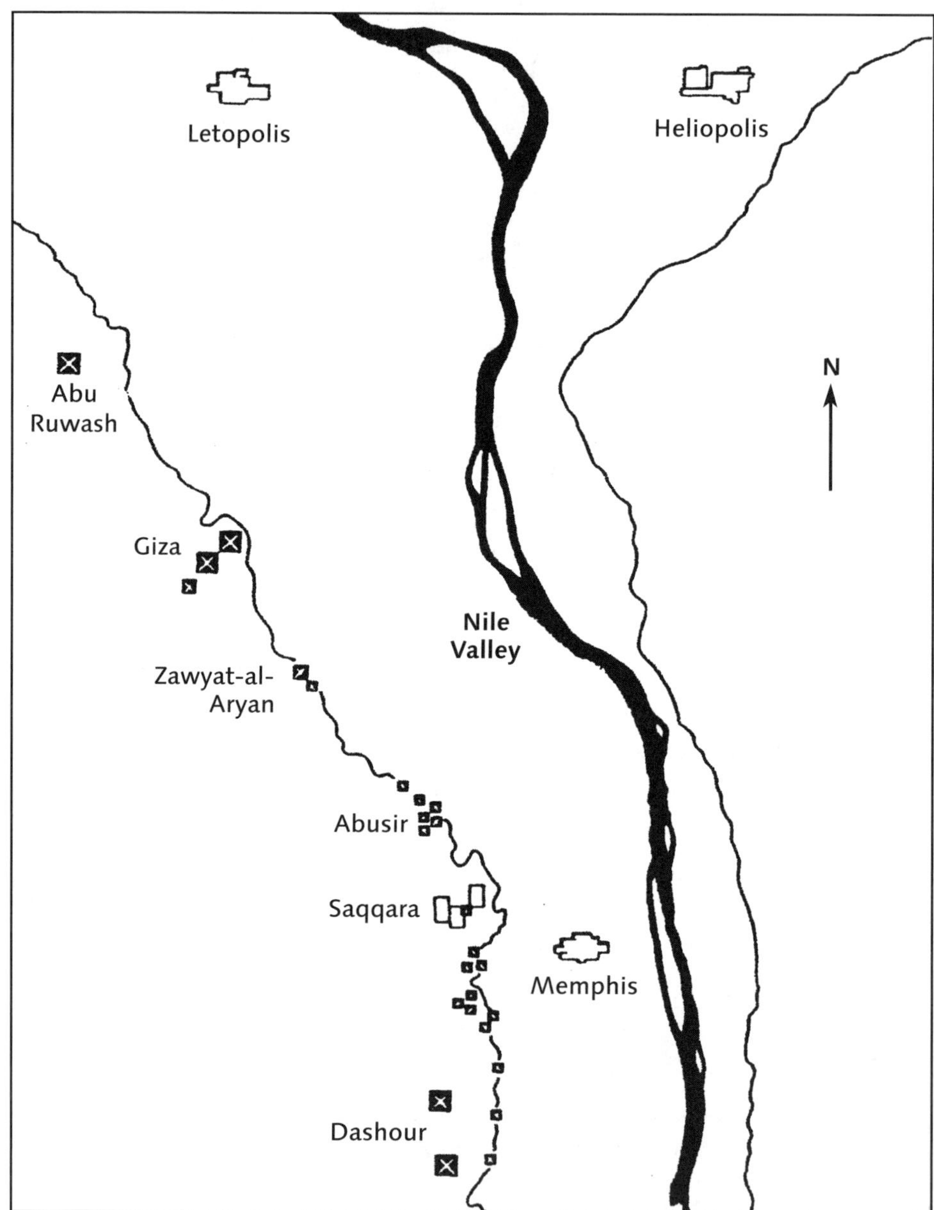

Figure 26. The pyramids of Lower Egypt are perched atop the edge of a limestone plateau located 'upwind' from the agricultural Nile Valley during the positively charged seasonal *khamsin* winds. The three pyramids of Giza line up aimed at the broadest swath of the valley. They were built here, right after this broad expanse 'downwind' began being farmed during an agricultural crisis. (After Robert Bauval & Adrian Gilbert. *The Orion Mystery.* Crown Publishers, New York, 1994)

a line atop the plateau where multiple aquifer layers surface,[45] in other words, atop an interfluve (Figure 26). The larger the base of the pyramid, the more ground current from the interfluve would be concentrated at the apex. The insulating Tura limestone cover would ensure that all charge rose toward the apex without leaking out around the sides. The positively charged *khamsin* wind would create a condition at the peak ideal for an electric brush discharge. The concentrated negative charge, accumulated from the drained limestone aquifers would connect with the oppositely charged wind. Not only would the top glow, it would release ionized nitrogen downwind to the farms.

The siting of the three Giza pyramids also suggests a reason for the selection of this particular plateau. It reminds us of the geological locations of Avebury and Silbury Hill (Figure 18). The Giza pyramids are all lined up along a tongue or 'peninsula' of the Mokkatam formation limestone. The building stones were quarried right where the quarries would act in a manner similar to a huge henge ditch. They cut into the peninsula's side, thereby narrowing it. This placement would further amplify the 'peninsula effect' discussed in Chapter Two. The result was to leave the giant Khufu pyramid with a base that covered most of the intact tongue of the Mokkatam formation[46]. The physics at work should function much like an undisturbed causeway in a henge, concentrating the ground current of the whole peninsula in this small piece of intact ground.

THE 'QUEEN'S' PYRAMIDS

All giant pyramids had a tiny, subsidiary pyramid built next to them, generally to the south or southeast. Archaeologists call them 'queen's' pyramids, assuming that their larger companions were the final resting places for the kings, but no one, in fact, has a clue as to what these structures were for. They have no chamber, no burial, no writing. Yet all major pyramids have at least one; Khufu's has several. For what reason could they be built?

One night, in a pitch black room, I noticed that my electric ion generator not only had the distinctive purple glow at the tip of the needle, indicating that air ionization was taking place, but also an extremely faint, second pinpoint of purple light, a couple of inches off to the side. This turned out to be a tiny, secondary needle we had never even noticed before. When I tried covering one of the needles with a finger, the purple glow on the other needle would wink out. It didn't matter whether we covered the major one or the minor one. Eliminating one always killed the other.

A phone call to the technical department of the manufacturer revealed that the smaller needle was a reduced, low-power version of the main needle. It turned out that the early, single needle ion generators had spread ions uniformly in all directions. This was a disadvantage to the homeowner who usually would place the unit on a table near a wall. Over time, the stream of ions hitting the wall tended to dirty it with all the dust particles they attract. The presence of a smaller low-power needle created an electric field that directed the ions from the major needle in one direction. Placing the small needle on the large one's south side would tend to direct the ions east and north. We wondered if this process could explain the placement of the pyramids.

The farmlands where the ionized nitrogen was needed lay east of the pyramids, stretching in a long line north and south. When you think about it, the Nile Valley is a tough target to hit, if what you are 'firing' has to be based outside the valley. Not enough loft, and you only reach the fields next to the pyramid. Too much loft and you overshoot the whole valley. Going due east only hits a narrow swath of the valley's length. Northeast and southeast would be the ideal trajectory, if you want to cover the maximum amount of farmland. The *khamsin* blows towards the north and northeast. As farmlands expanded north up the valley to the delta, the pyramids were situated where they would do most good. And if the subsidiary pyramids worked like the small needle in ion generators, they would also have tended to direct the *khamsin* flow off the main pyramid to the north and northeast.

Let us assume that Imhotep did want to build giant generators of ions and free nitrogen. Due to the concentration of positive atmospheric field lines at the peak, the benben atop would always be exposed to an airspace of higher positive charge than any area around it. During *khamsin* periods, this positive field in the air would be dramatically enhanced. This is also the time of the year when the negative electrical charge in the ground should be at its maximum. Therefore, the pyramids are sited where the electrical scenario is the most conducive to an outcome that the infertile fields would need most.

THE INNER CHAMBERS AND SHAFTS

One of the primary pyramid puzzles is the purpose of the chambers, such as the famous King's and Queen's chambers at Giza. Unlike the rest of the pyramid, they were lined with granite, the preferred rock of the Carnac chambers, which had to be transported a much greater distance. From the chambers, long, narrow shafts radiated out through the pyramid. The chambers were sealed shut, as were the shafts. While they came to be called airshafts, these foot-wide tunnels were blocked on both ends and so never transported air. However, both they and the chambers did have stagnant air trapped inside. This air would be ionized or electrified by the natural radioactivity from the granite, becoming veins of electrical charge lacing the inside of the pyramid. The air would also receive electrical input from the general electric current flowing from pyramid base to peak. It brings to mind the electrified 'veins' of the drains inside the Akapana pyramid of Tiwanaku (Chapter Five) and of the sculpted Olmec hill at San Lorenzo (Chapter One).

The shafts were an architect's or engineer's nightmare because the cut stones of the pyramid core had to have appropriate sections of the shafts cut through them at just the right angle and height. Try to imagine the complexity of designing this, cutting the many stones in the appropriate way, and then seeing that those particular stones got placed in exactly the right spots. It's enough to make us think that they must have had some substantial value to the builders, but just what exactly?

We wonder if these shafts may have acted like a car battery. After all, once your car is running, you don't need a battery until you stop. Then, if your alternator is not working perfectly, your engine may die because there was an interruption in the electrical supply to the spark plugs. During the *khamsin* wind, there had to be lulls between gusts and periods of weak wind on some days. Likewise, the negative electric charge from the ground into the base would vary with the time of the day, strongest at sunrise and perhaps sunset, weakest near mid-day, like ground currents anywhere.

Could the shafts have helped to maintain the electrical activity of the pyramid on a more constant basis during these periods of partial interruption or lulls? It is only speculation, but it at least has the virtue of accounting for why the builders would go to such extraordinary lengths to incorporate these comparatively tiny features.

THE MYSTERY OF THE SPHINX

It is our hypothesis that the Giza Plateau hosts the most enormous fertility generators ever created on this planet. As previously told, these pyramids are mysteriously absent of writing or even symbols that the architects might have used to hint at their purpose, giving at least some kind of context to these behemoths. There is one symbol among the pyramids, an icon so large it does justice to the blank pyramids.

The Sphinx is what archaeologists call a were-cat. It is half lion (the body), half human (the face). We realized that we had seen such icons elsewhere. A central symbol in the cultures of the Olmec and Mayan pyramid builders, and also of the Andean civilizations of Tiwanaku, and later, the Inca, was that of a were-cat, part jaguar and part man. Virtually all experts today agree that this were-cat was a symbol of fertility. It appeared several places near the stone fertility generators where we had measured energies. And now, the biggest were-cat in the world sits like a sentinel at Giza.

'LIGHT'

Our hypothesis would explain the siting of the pyramids, their timing, their design, their building materials, and why agricultural crisis framed their history. It would explain why they were built so incredibly large, despite the enormous cost to the society. It would explain the function of the small companion 'queen's' pyramids. Finally, it would explain why these gigantic stone structures were built with such amazing speed.

The inevitable conclusion is that the great Egyptian pyramids were fertility generators. They constituted the practical response of a practical people to a crisis that threatened their lives. This explains why the Egyptian people time and again willingly undertook construction projects that defy our imagination. The building of the pyramids was an agricultural people's desperate bid for survival. After all, fear of famine is a supreme motivator.

Let us finish by casting our mind's eye back 4,500 years. The pyramids glow a brilliant white in the desert sun by day and, perhaps, glow visibly from the peak by night. Even today, light phenomena are still seen, most commonly on the far side of the Giza pyramids, away from the bright lights of Cairo. They are so common there that the desert-dwelling bedouins have a special word for them.[47] And now consider that the very word the Romans gave us for these wonders of the world, *pyramid,* consists of the words *pyra,* 'fire,' and *mid,* 'middle,' thus 'fire in the middle.' The Egyptian word for pyramid is *takhat,* meaning simply 'light.'

Conclusion

"Only one thing can be stated with certainty about such structures as Stonehenge: The people who built them were much more intelligent than many who have written books about them. It cannot be too firmly stressed that the ancient architects — our ancestors of only two hundred *generations ago! — were men exactly like us. . . . there is no need to invoke any magical or mysterious powers, still less the intervention of any 'superior' beings, to explain their achievements."*

— Arthur C. Clarke[1]

It is our hope that this book may constitute one small step toward shedding some light on where we came from and where we are headed. We are gratified to know that today food production is once again being increased by a carefully controlled laboratory version of the energies that helped feed agricultural civilizations throughout history. We are also sobered to know just how capable our ancestors really were, and we bask in a certain pride about that.

A few warring camps seem to exist today in discussions of early civilizations. One is that these were simple people with simple tools and they just were not capable of doing much besides eking out a bare subsistence lifestyle. They are assumed to have possessed less knowledge than we do, and a more primitive mindset. It does not occur to the members of this camp to wonder if ancient peoples knew something we do not and that there may have been some practical purpose to it all that is outside our contemporary knowledge base. It seems odd that we should think this way when we blithely accept the existence of lost engineering knowledge. We know that in Peru the Inca were able to cut through enormous stones with such ease that they thought nothing of making towering walls out of a jigsaw of such stones, all custom cut and precisely fitted. We are unable to duplicate such stone masonry today and admit that the Inca knew something about stone work that we do not. And why should this knowledge be surprising from the best and brightest of the late Stone Age? Stone was their primary medium; it is not ours.

Consider the books and the documentaries still being issued every year, purporting to explain an exciting new theory on how the Egyptians built their colossal pyramids. We have been wondering for centuries and are still unsure. When it comes to the engineering, we are willing to grant that they knew practical things that we do not. Why then not grant the same benefit of the doubt as to the design and purpose of megalithic structures?

After all, theirs was a civilization of subsistence agriculture and ours is not. Those of us who live in the industrialized nations take it for granted that there will be enough food next year. If there is a massive drought striking our breadbasket regions, we will simply import the food we need next year. We focus on industry. While we may have some impressive cathedrals for religious use, by far our biggest structures are enormous, utilitarian hydro-electric dams, designed to produce the lifeblood of an industrial civilization: electricity. As we have seen, it looks like the biggest creations of our forebears were giant, utilitarian, megalithic structures designed to produce the lifeblood of an agricultural civilization: fertility.

Clearly, there was a high degree of technical competence in the areas we have studied in this book. There was also great understanding of physical forces that we are only now re-discovering. This should not surprise us. After all, pyramid builders had brains similar to ours. They did have a different knowledge base. Many things that we know they did not, and some things that they knew we do not. Does this mean they were far wiser than us? Not necessarily, just look at how they evolved into us after all, and how their societies often collapsed because of their own limitations. They were not superior to us, and neither were they dramatically inferior in intellect or ability.

These earth and stone fertility generators worked. You can test some of them yourself. Without fertilizer or crop rotation, the ancient engineers managed to not only feed a mushrooming population but to produce such a surplus as to grow rich enough to import luxury goods. When Pizarro reached Peru, he reported that the Inca were such successful food producers, even in their challenging terrain, that at any one time they had on hand enough stored grain to feed their population of millions for five years. In the year 1999, by contrast, when a late northern hemisphere harvest was finally reaped, the grain reserves of the modern world were down to a mere eighteen days.

Pizarro's firearms put an end to the Inca system. Many other ancient peoples, such as the Mayans, the Egyptians, and the Sumerians, were undone by decades-long droughts. So even with their successes, they lived with uncertainty. And it was to address that uncertainty that much of their ritual was directed. Traditionally, the ancients would not separate the physical from the non-physical, the soul from the land. We have just seen how tremendous effort was repeatedly taken to create an edifice that seems to us today to be imbued with an aura of ritual, and yet our experiments show that it is tapping natural energy in a way that can increase food production. Today's attitudes among most people of European cultural descent are shaped by the assumption that the human soul is qualitatively distinct from the landscape in which it dwells. We do not ourselves need to believe in the gods of our ancestors to appreciate that they lived closer to the land and were more aware of it. And we can rediscover some of their technology and profit by it, whether or not it may have once been wrapped in ritual.

Let us briefly review what we have seen in our journey through time and geography.

- The first high civilization of the Americas, the Olmec of Mexico, built earthen mounds and pyramids with well-engineered electrical characteristics. Villages with a mound enjoyed a better standard of living than otherwise identical villages nearby.

- The last Pyramid of the Olmec was the first of the Maya, The Lost World Pyramid of Tikal, Guatemala. At this pyramid, we measured tremendous surges of electrical ground current and airborne electric charge in the hours bracketing dawn. Local Mayan corn (maize) seed that we placed atop this pyramid during these electrical surges displayed tremendously increased germination and growth. The stronger the electrical activity on a given day, the greater the growth improvement. Seed was not improved when placed atop later pyramids, which are known to have been built as political monuments and which had no electrical activity. Mayans today still place their seed atop certain pyramids at dawn, and teach that plant growth is improved above the same kinds of geological structures that generate such electrical forces.

- Modern seed treatments using artificial versions of these natural forces are known to produce similar kinds of growth improvements, including higher yields. This has been confirmed dozens of times by universities and agricultural organizations. The degree of improvement is highest with the kinds of stressful growing conditions and relatively poor seed quality that typified the agriculture of the megalith builders.
- These kinds of energies are generated naturally everywhere every day by well-known forces of the Earth, but they are magnified at some locations by certain geological structures called conductivity discontinuities.
- Consistently, throughout history and all over the world, giant structures of earth and stone were built atop such conductivity discontinuities.
- Humans are sensitive to these forces, and some humans are capable of sensing them consistently. These forces have been used by pre-agricultural societies for vision quests. We have personally documented this use in a contemporary Native American setting.
- Mushrooms across Europe are substantially more prolific atop such geology. Ancient shamans frequently hunted mushrooms for their own vision quests. This is, therefore, a likely connection that would have been observed between these locations and improved plant growth.
- The Native American mound builders of North America were the cultural descendants of the Olmec and Maya. They located earthen mounds and pyramids atop magnetic anomalies and at conductivity discontinuities, where unusually strong electrical ground currents are generated – all confirmed in direct measurements by us. Seed left atop mounds during powerful energy surges have shown dramatic improvement in growth. These mounds can be visited today across most of the South and Midwest (see Appendix 1 for locations and directions).

- In the last of the mound-using societies of North America, farmers would not dream of planting their seed without first taking it to the top of a mound for "certain blessings."

- The chief role of much Native American religion was to help the crops grow better.

- Rock chambers throughout the Northeastern United States were built atop magnetic anomalies at conductivity discontinuities just like the mounds. We spent years measuring electric charge separation inside these chambers and improved growth from maize, wheat, and bean seeds left inside during days with electric charge, as compared to control seeds left outside. Results met scientific standards for high statistical significance. The biochemical changes inside this seed have been found to be the same kinds of biochemical changes found inside seed treated with today's modern laboratory version of these energies.

- Nine samples of corn (maize) seed of a variety grown by Native Americans from 700 AD onward was placed inside thee different rock chambers for varying amounts of time, then planted and grown out organically. When compared to nine plots of control corn kept outside the chamber or at home in the laboratory, the chamber-exposed seed produced, on average, double or triple the yields of the plots of control seed.

- Centuries before the Inca, a mysterious civilization on the Peruvian-Bolivian border built rock chambers of the same type found in the northeastern U.S., as well as sophisticated pyramids with electrical properties. Accounts by early Spanish colonizers relate that the farmers of this civilization somehow grew triple the amount of food that today's farmers can grow at the same location, in the harsh Andean *altiplano.*

- Many thousands of standing stones around Carnac, France, were placed solely at the border of zones of differing magnetic strength and seismic activity. Numerous series of enormous stone chambers were placed one after another precisely above invisible fault lines.

Measurements of the type we have taken at megalithic structures were carefully performed by a Belgian engineer over a thirty-year period, with results strikingly similar to ours, showing magnetic and electric variations of similar magnitude. All these structures were erected only after a food crisis was on hand, starting 6,700 years ago – more than 2,000 years before the existence of the pyramids of Egypt.

- In England, we measured the effectiveness of henges at concentrating electrical ground currents by their placement, their design, and their construction. Sited on the same types of geology we have seen elsewhere, we saw numerous archaeological finds where seed had been brought to these henges. Stonehenge was finally abandoned, after 2,000 years of use, only when fertilizer and crop rotation were introduced, thereby allowing farmers to multiply their yields without carrying seed to henges.
- While the situation in Egypt was somewhat unique, we still saw all the same factors in place that we have seen elsewhere: famine, electrically active geological and atmospheric conditions, a harnessing of these energies in a manner that could have helped relieve the famine, and a chronology of building successive pyramids alongside and "upwind" of the new farming regions as they opened up. We see how these energies could first have come to the attention of Imhotep, the first architect and chief engineer of the giant pyramids, in the temple at Heliopolis where he was also head priest. And we saw how those energies have manifested themselves at the same location in modern times before tens of thousands of witnesses in the form of glowing balls of light resulting from brush discharge. We began the chapter with the account of an early scientific pioneer of electricity experiencing powerful electric charge atop the great pyramid at Giza. And, finally, we saw how every aspect of the design of the giant pyramids, from the building materials for each part, to the fitting of the casement stones, to the 'airshafts' and chambers, might well have functioned to efficiently collect, concentrate, and harness these energies to ionize air and produce airborne nitrogen fertilizer to the farms downwind at the same time of year and time in history when it was both possible and urgently needed.

Anyone who wants to can obtain instruments similar to those we used (see Appendix 2) and visit any of the many dozens of sites in the U.S., Canada, or England (see Appendix 1), or elsewhere around the world. Anyone can also repeat the seed experiments shown here (see Appendix 2). Whether or not you test our findings for yourself, we hope that you may come to share this new outlook on our ancestors and help resurrect precious knowledge that has been lost to us for thousands of years.

We do not ourselves need to believe in the gods of our forebears to appreciate that they lived closer to the land and were more aware of it. By contrast, we find ourselves a bit embarrassed that today we can live surrounded by some of the forces they tamed, and yet be completely oblivious to them. As we have seen, before there was agriculture, shamans tapped into these same forces to provide life-changing experiences for their people. Certainly anyone who had a base of such experience must have moved through the natural world with a different attitude than today's casual picnicker in a park. Are such attitudes now obsolete in the modern world? Are they of no use to us any longer? Look around your own daily world and decide for yourself. For our ancestors, the land was alive. It can be for us, too.

References and Notes

CHAPTER 1: THE LOST WORLD

[1]J. Zlotnicki & J.L. Le Mouel. 'Possible electrokinetic origin of large magnetic variations at La Fournaise volcano.' *Nature,* vol. 343, 15 Feb 1990, pp. 633-635

[2]Ralph Markson & Richard Nelson. 'Mountain-Peak Potential-Gradient Measurements and the Andes Glow.' *Weather,* vol. 25, 1970, p. 357

[3]Michael D. Coe & Richard A. Diehl. *In the Land of the Olmec.* University of Texas Press, Austin, 1980

[4]Coe & Diehl, op. cit., p. 28

[5]Coe & Diehl, op. cit., p. 30

[6]Anonymous, 'The Amateur Scientist,' *Scientific American,* vol. 175, 1960

[7]William Thompson (aka Lord Kelvin). *Reprints of Papers on Electrostatics and Magnetism.* Macmillan, London, 1872, pp. 319-325

[8]Coe & Diehl, op. cit., p. 379

[9]Coe & Diehl, op. cit., vol. 2, pp. 151-52

[10]William F. Rust & Robert J. Sharer. 'Olmec Settlement Data from La Venta, Tabasco, Mexico.' *Science,* vol. 242, 1988, pp. 102-104

[11]Robert H. Fuson. 'The Orientation of Mayan Ceremonial Centers.' *ANNALS,* Assoc. of American Geographers, vol. 59, 1969, pp. 508-510

[12]John B. Carlson. 'Lodestone Compass: Chinese or Olmec Primacy?' *Science*, vol. 189, 1975, pp. 752-760

[13]V.H. Malmstrom. 'Knowledge of Magnetism in Pre-Columbian Mesoamerica.' *Nature*, vol. 259 (5542) , 1976, pp. 390-391

[14]V.H. Malmstrom & Paul E. Dunn. 'Pre-Columbian Magnetic Sculptures in Western Guatemala,' 1979; summarized on the Science page of *Time Magazine,* September 3, 1979, under the title 'Fat Boys'

[15]Coe & Diehl, op. cit.

CHAPTER 2: HARNESSING NATURE'S ELECTROMAGNETIC ENERGY

[1]Tsuneji Rikitake & Yoshimari Honkura. *Solid Earth Geomagnetism.* Terra Scientific Publishing, Tokyo, 1985, pp. 296-325

[2]Rikitake & Honkura, op. cit., p. 296

CHAPTER 3: HOW DID THEY KNOW?

[1]David P. Barron & Sharon Mason. *The Greater Gungywamp. North Groton, CT. A Guidebook.* Gungywamp Society, Noank, Connecticut, 1994

[2]S. Subrahmanyam, P.V.S. Narayan & T.M. Srinivasan. 'Effect of Magnetic Micropulsations on the Biological Systems – a Bioenvironmental Study.' *International Journal of Biometeorology,* vol. 29, no. 3, 1985, pp. 293-305

[3]M.A. Persinger & C. Psych. 'Sudden unexpected deaths in epileptics following sudden, intense, increases in geomagnetic activity: prevalence of effect and potential mechanisms,' *International Journal of Biometeorology,* vol. 38, 1995, p. 180

[4]Hans-Dieter Betz. 'Unconventional Water Detection.' *Journal of Scienctific Exploration*, vol. 9, No. 2, 1995

[5]Hans-Dieter Betz et al. 'Dowsing Reviewed – the effect persists.' *Naturwissenschaften*, vol. 83, 1996, pp. 272-275

[6]*Scientific American,* vol. 96, 1907, p. 90

[7]Andrei Apostol. 'North American Indian Effigy Mounds: An Enigma at the Frontier of Archaeology and Geology.' *Journal of Scientific Exploration,* vol. 9, No 4, Article 3, 1996

[8]Marcia Barinaga. 'Giving Personal Magnetism a Whole New Meaning.' *Science,* vol. 256, 15 May 1992, p. 967

[9]Robert O. Becker & Gary Seldon. *The Body Electric.* William Morrow, 1985

[10]Michael E. Long. 'Secrets of Animal Navigation.' *National Geographic Magazine,* June 1991, pp. 70-99

[11]Winifred Gallagher. *The Power of Place – How Our Surroundings Shape Our Thoughts, Emotions, and Actions.* Poseidon Press/Simon & Schuster Inc., 1994

[12]William K. Powers. *Yuwipi. Vision and Experience in Oglala Ritual.* University of Nebraska Press, Lincoln, 1982

[13]Persinger & Psych, op. cit.

[14]BBC News: www.bbc.co.uk/science/horizon/2003/godonbrain.shtml

[15]M.A. Persinger. 'Increased geomagnetic activity and the occurrence of bereavement hallucinations: evidence for melatonin-mediated microseizuring in the temporal lobe.' *Neuroscience Letters*, vol. 88, 1988, pp. 271-284

[16]Jiri Dvorak, personal communication

[17]Peter T. Furst (ed.). *Flesh of the Gods. The Ritual Use of Hallucinogens.* Waveland Press, 1972

[18]Joseph Campbell. *The Way of the Animal Powers. Historical Atlas of World Mythology,* vol. 1, p. 173. Alfred van der Marck Editions, 1983

[19]Starr Fuentes, personal communication

CHAPTER 4: RETURN TO THE LOST WORLD

[1]Alfonso Caso. *The Aztecs: People of the Sun.* University of Oklahoma Press, Norman, 1958, p. 41

[2]Karen Bassie-Sweet. *At the Edge of the World: Caves and the Late Classic Maya World View.* University of Oklahoma Press, Norman, 1996, pp. 111-131

[3]Henry C. Mercer. *The Hill Caves of the Maya.* University of Oklahoma, Norman, 1896, p. xxviii

[4]Robert Redfield & Alfonso Villa Rojas. *Chan Kom: A Maya Village.* University of Chicago Press, 1962, p. 28

CHAPTER 5: BEFORE THE INCA

[1]Juan Schobinger. *The First Americans.* W.B. Eerdmans Publishing Co., Grand Rapids, 1994

[2]Richard L. Burger. 'The Prehistoric Occupation of Chavin de Huantar, Peru.' *Anthropology,* vol. 14, 1984, p. 248, University of California Press, Berkeley

[3]Schobinger, op. cit.

[4]Burger, op. cit., p. 219

[5]Olivia Vlahos. *New World Beginnings: Indian Cultures in The Americas.* Viking Press, New York, 1970, p. 261

[6]Alan L. Kolata. *The Tiwanaku: Portrait of an Andean Civilization.* Blackwell Press, Cambridge, Massachusetts, & Oxford, England, 1993

[7]Schobinger, op. cit., p. 180

[8]*Lost Crops of the Incas.* Report of an Ad Hoc Panel of the Advisory Committee on Technology Innovation Board on Science and Technology for International Development, National Research Council. National Academy Press, Washington, D.C., 1989, pp. 8-9

[9]Alan L. Kolata. *Valley of the Spirits: A Journey into the Lost Realm of the Aymara.* John Wiley & Sons, New York, 1996, p. 176

[10]Kolata, 1996, op. cit., p. 8

[11]Max Mille & Carlos Ponce Sangines. *Las andesitas de Tiwanaku.* Academia Nacional de Ciencias de Bolivia, Publication no. 18, La Paz, 1968

[12]Kolata, 1996, op. cit., pp. 139-140

[13]Arthur Posnansky. *The Cradle of American Man.* J.J. Augustin Co., New York, 1945, p. 74

[14]Kolata, 1996, op. cit., p. 52

[15]Kolata, 1996, op. cit., p. 279

[16]Hans van den Berg. *La tierra no da asi no mas: los ritos agricolas en la religion de los aymara-cristianos.* Hisbol, La Paz, 1990, cited in: Kolata, 1996, op. cit., p. 249

[17]Kolata, 1996, op. cit., p. 249

[18]Kolata, 1993, op. cit., p. 111

[19]Posnansky, op. cit., p. 74

[20]Posnansky, op. cit., p. 73

[21]Kolata, 1993, op. cit., p. 131

[22]Kolata, 1993, op. cit., p. 141

[23]Stig Ryden. *Archaeological Researches in the Highlands of Bolivia.* Elanders Boktryckeri Aktiebolag, Goteborg, Sweden, 1947, pp. 370-388

[24]Adolph F. Bandelier. *The Islands of Titicaca and Koati.* The Hispanic Society of America, 1910, p. 218

[25]Michael Coe, Dean Snow & Elizabeth Benson. *Atlas of Ancient America.* Facts on File, New York, 1986, p. 198

[26]Bandelier, op. cit., p. 218

[27]Bandelier, op. cit., p. 228

[28]G. Courty. *Explorations Geologiques dans l'Amerique du Sud.* Imprimerie Nationale, Paris, 1907, pp. 65-68

[29]Bandelier, op. cit., pp. 177-186

[30]Bandelier, op. cit., p. 221

[31]Bandelier, op. cit., p. 222

[32]Johan Reinhard. 'Sacred Peaks of the Andes.' *National Geographic Magazine,* vol. 181, no. 3, 1992, p. 103

[33]Kolata, 1996, op. cit.

[34]Coe et al., op. cit., p. 189

CHAPTER 6: ROCK CHAMBERS OF NEW ENGLAND

[1]Phil Imbrogno, personal communication

[2]Phil Imbrogno, personal communication

[3]Michael J. Rycroft. 'Active Experiments in Space Plasma.' *Nature,* vol. 287, 1980, p. 11

[4]Timothy Champion, Clive Gamble, & Stephen Shennan. *Prehistoric Europe.* Academic Press, London, 1984, p. 118

[5]C.L. Wiedenmann, *Vestigia Newsletter,* vol. 2, 1977, p. 1

[6]W.C. Levengood. 'Redox-responsive electrodes applied during plant morphogenesis.' *Bioelectrochemistry and Bioenergetics,* vol. 19, 1988, pp. 461-476

[7]W.C. Levengood, Lab Report, Pinelandia Biophysical Laboratory, Michigan

[8]Salvatore Trento. *The Search for Lost America.* Contemporary Books, Chicago, 1978, pp. 147-190

[9]Barry Fell. *Bronze Age America*. Little, Brown, & Co., New York, 1982, p. 53

[10]David P. Barron & Sharon Mason. *The Greater Gungywamp, North Croton, CT. A Guidebook*. Gungywamp Society, Noank, Connecticut, 1994, p. 3

[11]Alex Patterson, personal communication

[12]Joseph Campbell. *The Mythic Image*. Princeton University Press, 1981

[13]Karen Bassie-Sweet. *At the Edge of the World: Caves and the Late Classic Maya World View*. University of Oklahoma Press, Norman, 1996, p. 110

[14]Henry C. Mercer. *The Hill Caves of the Maya*. University of Oklahoma, Norman, 1896, pp. xxi-xxvii

[15]Robert Redfield & Alfonso Villa Rojas. *Chan Kom: A Maya Village*. University of Chicago Press, 1962, p. 28

[16]Bassie-Sweet, 1996, op. cit., pp. 111-131

[17]Bassie-Sweet, 1996, op. cit., p. 63

[18]Joseph Campbell. *The Way of the Animal Powers. Historical Atlas of World Mythology*, vol. 1, 1983, p. 17. Alfred van der Marck Editions

[19]E.S. Thompson, 1970, p. 351, cited in Karen Bassie-Sweet. *From the Mouth of the Dark Cave*. University of Oklahoma Press, Norman, 1991, p. 86

CHAPTER 7: MYSTERY MOUNDS OF NORTH AMERICA

[1]Ake Hultkrantz. *The Religions of the American Indians*. University of California Press, Berkeley, 1979

[2]Gene S. Stuart. *America's Ancient Cities*. National Geographic Society, Washington, D.C., 1988, p. 26

[3]Stuart, op. cit., p. 25

[4]Reuben Gold Thwaites (ed.). *The Jesuit Relations and Allied Documents*, vol. IXVIII, doc. CCIII. In 73 volumes, selected and edited in one volume by Edna Kenson. Albert & Charles Boni, New York, 1925

[5]*Archaeology*, vol. 49, no. 6, 1996, p. 16

[6]Stuart, op. cit., p. 36

[7]Cahokia Mounds State Park museum exhibit, Collinsville, Illinois

[8]Karen Bassie-Sweet. *From the Mouth of the Dark Cave.* University of Oklahoma Press, Norman, 1991, p. 86

[9]Karen Bassie-Sweet. *At the Edge of the World: Caves and the Late Classic Maya World View.* University of Oklahoma Press, Norman, 1996, pp. 81, 84

[10]Robert C. Glotzhober & Bradley T. Lepper. *Serpent Mound: Ohio's Enigmatic Effigy Mound.* Ohio Historical Society, Columbus, 1994

[11]Hopewell Culture National Historical Park museum exhibit, Chillicothe, Ohio

[12]Dr. Bradley Lepper, personal communication

[13]J.P. Cull & D.H. Tucker. 'Telluric currents and magnetic anomalies.' *Geophysical Research Letters,* vol. 13, no. 9, 1986, pp. 941-944

[14]Cahokia Mounds State Park Museum, Collinsville, Illinois

[15]Thomas Emerson, Illinois Historic Preservation Agency, interviewed in the video *Cahokia Mounds: Ancient Metropolis,* Cahokia Mounds Museum Society

[16]Stuart, op. cit., p. 36

[17]*Science,* vol. 272, 1996, p. 351

[18]P.F.X. de Charlevoix. *Histoire et Description Generale de la Nouvelle France,* vol. VI, p. 183, Paris, 1744

[19]Robert Neitzel. *Archaeology of the Fatherland Site: The Grand Village of the Natchez.* Anthropological Papers of the American Museum of Natural History, vol. 51, New York, 1955, pp. 69, 71

[20]Patricia Dillon Woods. *French-Indian Relations on the Southern Frontier, 1699-1762.* Studies in American History and Culture, no. 18, University of Michigan Research Press, Ann Arbor, 1980, p. 4

[21]Neitzel, op. cit., p. 70

[22]Neitzel, op. cit., p. 71

[23]Woods, op. cit., p. 4

[24]John R. Swanton. 'Source Material for the Social and Ceremonial Life of the Choctaw Indians.' *Bureau of American Ethnology,* Bulletin 103, 1931, p. 13. U.S. Government Printing Office, Washington, D.C.

[25]Choctaw Tribe's Website

[26]Woods, op. cit., p. 4

[27]Woods, op. cit., p. 4

[28]Reuben Gold Thwaites, op. cit., vol. 15, p. 157

[29]Reuben Gold Thwaites, op. cit., vol. 15, p. 153, vol. 19, p. 133

[30]Lawrence Davis-Hollander, personal communication

[31]Ulysses Prentiss Hendrick. *A History of Agriculture in New York State*. New York State Agriculture Society, 1933, p. 25

[32]Addison J. Throop. *Mound Builders of Illinois*. Call Printing Co., E. St. Louis, IL, 1928, p. 69

[33]Stuart, op. cit., pp. 38-44

[34]Guy Prentice, quoted by Stuart, op. cit., p. 36

[35]Cyrus Thomas. *The Cherokees in Pre-Columbian Times.* Hodges Publishers, New York, 1890

[36]Bassie-Sweet, 1996, op. cit., p. 63

[37]Henry C. Mercer. *The Hill Caves of the Maya.* University of Oklahoma, Norman, 1896, pp. 142-143

[38]Bassie-Sweet, 1996, op. cit., pp. 66-69

[39]Doris Heyden. *The Eagle, the Cactus, and the Rock.* BAR International Series, 484, 1989, pp. 63-64

[40]Eduardo Matos Moctezuma. *The Great Temple of the Aztecs.* Thames & Hudson, London, 1988, p. 142

[41]Brian M. Fagan. *Kingdoms of Gold, Kingdoms of Jade: The Americas Before Columbus.* Thames & Hudson, London, 1991, p. 32

[42]*Encyclopedia Britannica,* vol. 26, 1985, p. 17

[43]Alan R. Sandstrom. *Corn Is Our Blood.* University of Oklahoma Press, Norman, 1991, p. 288

[44]Harold E. Driver. *Indians of North America.* 2nd edition, University of Chicago Press, 1969, p. 396

[45]Ake Hultkrantz, op. cit.

CHAPTER 8: THE FIRST MEGALITHS

[1]Pierre Mereaux. *Carnac – des pierres pour les vivants.* Nature et Bretagne, Spezet, Bretagne, 1992 (excerpts translated by Roslyn Strong)

[2]Peter Bogucki. 'The Spread of Early Farming in Europe.' *American Scientist,* vol. 84, 1996, pp. 242-245

[3]*Past Worlds. The Times Atlas of Archaeology.* Crescent Books, Avenel, New Jersey, 1995, p. 86

[4]Chris Scarre, Roy Switsur & Jean-Pierre Mohen. 'New radiocarbon dates from Bougon and the chronology of French passage graves.' *Antiquity,* vol. 67, No. 257, 1993, pp. 856-859

[5]Roger Joussaume. *Dolmens for the Dead. Megalith-Building throughout the World.* Cornell University Press, Ithaca, New York, 1988, p. 23

[6]Mark Patton. *Statements in Stone: Monuments and Society in Neolithic Brittany.* Routledge, London, 1993, pp. 65-66

[7]Patton, op. cit., p. 65

[8]Scarre, Switsur & Mohen, op. cit.

[9]Colin Burgess. *The Age of Stonehenge.* J.M. Dent & Sons Ltd., London, 1980, p. 52

[10]Michael Balter. 'New Look at Neolithic Sites Reveals Complex Societies.' *Science,* vol. 262, 8 Oct 1993, p. 179

[11]Jill L. Furst. *Codex Vinobonensis Mexicanus I: A Commentary.* Institute for Mesoamerican Studies, University of Albany, pub. no. 4, 1978, p. 318

[12]James Mellart. *Catal Huyuk: A Neolithic Town in Anatolia.* McGraw-Hill, New York, 1967, pp. 17-26

[13]Balter, op. cit., pp. 179-180

[14]Balter, op. cit., p. 180

[15]Mereaux, op. cit., 1992, p. 47

[16]*Science,* vol. 270, 11/10/95, p. 911

[17]Brady & Rowell. 'Laborarory Investigation of the Electrodynamics of Rock Fracture.' *Nature,* vol. 321, 1986, p. 490

[18]Pierre Mereaux. *Carnac – une porte vers l'inconnu.* Editions Robert Laffont, Paris, 1981, p. 261

[19]Tsuneji Rikitake & Yoshimari Honkura. *Solid Earth Geomagnetism.* Terra Scientific Publishing, Tokyo, 1985, pp. 296-320

[20]Patton, op. cit., p. 107

[21]Don Wallace. *New York Times,* 7/14/96, sec. 5, p. 14

[22]Mereaux, 1992, op. cit.

[23]Mereaux, 1992, op. cit., p. 47

[24]Mereaux, 1981, op. cit., p. 261

[25]Aubrey Burl. *From Carnac to Callanish: the prehistoric stone rows and avenues of Britain, Ireland, and Brittany.* Yale University Press, 1993, p. 49

[26]Mereaux, 1981, op. cit., p. 261

CHAPTER 9: THE HENGES OF SOUTHERN ENGLAND

[1]Colin Burgess. *The Age of Stonehenge.* J.M. Dent & Sons Ltd., London, 1980, p. 234

[2]Burgess, op. cit., pp. 235-236

[3]Aubrey Burl. *From Carnac to Callanish: the prehistoric stone rows and avenues of Britain, Ireland, and Brittany.* Yale University Press, 1993

[4]Rodney Castledon. *The Stonehenge People: An Exploration of Life in Neolithic Britain 4700-2000 BC.* Routledge & Kegan Paul, London, 1987, pp. 48, 54, 139, 215

[5]Julian Thomas. *Rethinking the Neolithic.* Cambridge University Press, 1991, p. 33

[6]Aubrey Burl. *Prehistoric Avebury.* Yale University Press, New Haven, 1979, p. 25

[7]Caroline Malone. *The English Heritage Book of Avebury.* Batsford, London, 1989, p. 79

[8]Thomas, op. cit., p. 33

[9]Malone, op. cit.

[10]Michael Dames. *The Silbury Treasure. The Great Goddess Rediscovered.* Thames &

Hudson, London, 1976, p. 52

[11]Burl, 1979, op. cit., p. 24

[12]Thomas, op. cit., p. 33

[13]Francis Pryor. *Excavations at Fengate, Peterborough, England: the 4th Report.* Northamptonshire Archaeological Society, Monograph no. 2, 1984, p. 206

[14]Thomas, op. cit.

[15]Burl, 1979, op. cit., pp. 40-44

[16]H. Mizutani, T. Ishido, T. Yokokura, & S. Ohnishi. 'Electrokinetic Phenomena Associated with Earthquakes.' *Geophysical Research Letters,* vol. 3, no. 7, 1976, pp. 365-369

[17]Francois M.M. Morel. *Principles of Aquatic Chemistry.* John Wiley & Sons, New York, 1983

[18]L.N. Plummer & E. Busenberg. 'The solubilities of calcite, aragonite and vaterite in CO_2-H_2O solutions between 0 and 90°C, and an evaluation of the aqueous model for the system $CaCO_3$-CO_2-H_2O.' *Geochimica et Cosmochimica Acta,* vol. 46, 1982, pp 1011-1040

[19]J.P. Cull & D.H. Tucker. 'Telluric currents and magnetic anomalies.' *Geophysical Research Letters,* vol. 13, no. 9, 1986, pp. 941-944

[20]R.J. Martin et al. 'The Effect of Fluid Flow on the Magnetic Field in Low Porosity Crystalline Rock,' *Geophysical Research Letters*, vol. 9, No. 12, pp. 1301-1304, 1982

[21]U.K. Government Survey, www.nwl.ac.uk/ih/nrfa/yb/yb2001/groundwater.html

[22]Martin et al., op. cit.

[23]Caroline Malone. *Prehistoric Monuments of Avebury, Wiltshire.* English Heritage, London, 1990, p. 4

[24]*American Scientist*, vol. 58, 1970, p. 272

[25]Susan Michaels. *Sightings.* Simon & Schuster, New York, 1996, p. 99

[26]Burl, 1979, op. cit., p. 113

[27]Castledon, op. cit., p. 237

[28]Dames, op. cit.

[29]V.P. Hessler & E.M. Wescott. 'Correlation between Earth-Current and Geomagnetic Disturbance.' *Nature,* vol. 184, Aug 22, 1959, p. 627

[30]Burl, 1979, op. cit.

[31]Aubrey Burl. *The Stone Circles of the British Isles.* Yale University Press, 1976, p. 324

[32]Malone, op. cit.

[33]Some may claim that we did *not* measure the magnetism of the stones themselves, but rather of the iron bars in the concrete, used when the stones were re-erected in the 1930s. At first, we thought that these bars had been magnetized in a north-south position, when cast at the factory, and what we were measuring was merely the magnetic field of the bars, nothing to do with the stones. But photos in Avebury Museum of the renovation of the complex clearly showed that each stone was surrounded by a nest of interwoven iron bars, pointing in all directions, some even perpendicular to the axis of the magnetic field we were now measuring. Yet our magnetometer recorded a single coherent field associated with each stone. Clearly, the iron bars had, over time, received an induced magnetization from the stones. You may have experienced something similar if you have left a nail or a screwdriver stored next to a magnet. Next time you want to use it, it has become magnetic. Before, the magnetic north and south poles of its iron molecules would point randomly and cancel each other out. Now, its poles would have been 'coerced' to point the same way as the magnet. Over the past seventy years, the iron bars of Avebury have slowly had their magnetic north and south poles reoriented by the fixed field of the stone. This means that the stones themselves were magnetic all along, but less than one hundred gammas in strength, too weak for us to have detected with our magnetometer. Our great luck was that the iron bars had been added, otherwise we would never have discovered the pattern of magnetization. As we know, though, shamans and some other individuals can detect much smaller differences (see Chapter Three). People like these, it seems, built Avebury.

[34]Burl, 1979, op. cit., p. 186

[35]Paul Devereux. *Places of Power. Secret Energies at Ancient Sites: A Guide to Observed or Measured Phenomena.* Blandford, London, 1990, pp. 135-136

[36]Jason Goodwin. 'The Survivors of Avebury,' *The New York Times,* 7/14/96,

Sec. 5, p. 15

[37]Paul Devereux, John Steele & David Kubrin. *Earthmind. A Modern Adventure in Ancient Wisdom.* Harper & Row, New York, 1989, p. 104

[38]Otto T. & Dorothy J. Solbrig. *So Shall You Reap.* Island Press, Washington, D.C., 1994

[39]Castledon, op. cit., pp. 214, 218

[40]R.P. Harrison et al. 'Helicopter observations of very low frequency radio waves over certain mountains and shore lines.' *Journal of Atmospheric and Terrestrial Physics,* vol. 33, 1971, pp. 101-106

[41]Burl, 1979, op. cit., p. 113

[42]R.M.J. Cleal, K.E. Walker & R. Montague. *Stonehenge and Its Landscape.* English Heritage, London, 1995, p. 499

[43]Burgess, op. cit.

[44]Burl, 1979, op. cit., p. 325

[45]J.D.A. Piper. *Paleomagnetism and the Continental Crust.* John Wiley & Sons, New York, 1987, pp. 6-7

[46]R.J.C. Atkinson. *Stonehenge and Neighbouring Monuments.* English Heritage, London, 1987, p. 15

[47]Atkinson, op. cit., p. 17

[48]Cleal et al., op. cit., p. 509

[49]Timothy Champion, Clive Gamble, & Stephen Shennan. *Prehistoric Europe.* Academic Press, London, 1984

[50]Malone, op. cit.

[51]Atkinson, op. cit., p. 27

[52]Atkinson, op. cit.

[53]Barry Cunliffe (ed.). *The Oxford Illustrated History of Europe.* Oxford University Press, 1994, p. 315

CHAPTER 10: PULSE OF THE PYRAMID

[1]Rushdi Said. *The River Nile: Geology, Hydrology, and Utilization.* Pergamon Press,

Oxford, 1993, p. 127

[2]Peter Tompkins. *Secrets of the Great Pyramid.* Harper & Row, New York, 1971

[3]*Past Worlds. The Times Atlas of Archaeology.* Crescent Books, Avenel, New Jersey, 1995, p. 94

[4]Otto T. & Dorothy J. Solbrig. *So Shall You Reap.* Island Press, Washington, D.C., 1994

[5]John Waterbury. *Hydropolitics of the Nile Valley.* Syracuse University Press, 1979, p. 148, Table 12

[6]*Past Worlds. The Times Atlas of Archaeology,* op. cit., p. 129

[7]*Past Worlds. The Times Atlas of Archaeology,* op. cit., p. 129

[8]Jaromir Malek (ed.). *Egypt: Cradle of Civilization.* University of Oklahoma Press, Norman, 1993, p. 17

[9]Said, op. cit., p. 128

[10]Solbrig, op. cit., pp. 116-117

[11]Solbrig, op. cit.

[12]Jacquetta Hawkes. *The First Great Civilizations.* Alfred A. Knopf, New York, 1973, p. 340

[13]I.E.S. Edwards. *The Pyramids of Egypt.* Penguin Books, London, 1993, p. 34

[14]Robert Bauval & Adrian Gilbert. *The Orion Mystery.* Crown Publishers, New York, 1994, p. 26

[15]R.T.R. Clark. *Myth and Symbol in Ancient Egypt.* Thames & Hudson, London, 1978, pp. 37-61

[16]Clark, op. cit., p. 246

[17]John Baines. *Orientalia,* vol. 39, 1970, pp. 389-395

[18]E.A. Wallis-Budge. *Cleopatra's Needles.* London, 1926

[19]Bauval & Gilbert, op. cit., pp. 201-02

[20]G.A. Wainwright. 'Some Aspects of Amun.' *Journal of Egyptian Archaeology,* vol. 18, 1933, pp. 3-15

[21]N. Robinson & F.S. Dirnfeld. 'The ionization state of the atmosphere as a function of meteorological elements and of various sources of ions.' *International*

Journal of Biometeorology, vol. 6, 1963, pp. 101-110

[22]I.N. Malysheva. 'Meteorological and physiological significance of atmospheric ionization.' *Vestnik. Akad. Med. Nauk.,* SSSR, vol. 19, 1964, pp. 83-89 (abstract)

[23]M. Assael, Y. Pfeifer & F.G. Sulman. 'Influence of Artificial Air Ionization on the Human Electroencephalogram.' *International Journal of Biometeorology,* vol. 18, 1974, pp. 306-312

[24]A.L. Chizevskii, in: *The Earth in the Universe* (ed. V.V. Fedynskii), published for NASA by the Israel Program for Scientific Translation, Jerusalem, 1968, pp. 350-351

[25]*U.S. News & World Report,* vol. 67, 12/29/69, p. 66

[26]*Psychology Today,* Jan/Feb 1993, p. 64

[27]*The New York Times,* May 5, 1968, p. 71

[28]*The New York Times,* May 21, 1968, p. 16

[29]J.S. Derr & M.A. Persinger. 'Geophysical Variables and Behavior: LIV. Zeitoun (Egypt) Apparitions of the Virgin Mary as Tectonic Strain-induced Luminosities.' *Perceptual and Motor Skills,* vol. 68, 1989, pp. 123-128

[30]Kruger, A.P. and Yamaguchi, F.M. 'Electroculture of tomato plants in a commercial hydroponics greenhouse,' *Journal of Biological Physics,* 11: pp. 5-10 (1983)

[31]Ralph Markson & Richard Nelson. 'Mountain-Peak Potential-Gradient Measurements and the Andes Glow.' *Weather,* vol. 25, 1970, p. 357

[32]*Geological Map of Greater Cairo Area.* Geological Survey of Egypt, Cairo, 1983

[33]M.M. Hassaan, A.M. Hassanein, M.H. Mohammad, & S.A. Abd El Moneim. 'Lithostratigraphical and Microfacies Studies on the Mokattam and Maadi Formations, Gabal Mokattam Area, Egypt.' *Annals of the geological survey of Egypt,* vol. XVI, 1986-89, pp. 229-233

[34]John A. Wilson. *The Burden of Egypt: An Interpretation of Ancient Egyptian Culture.* University of Chicago Press, 1951, p. 55

[35]*Encyclopedia Britannica,* vol. 18, p. 151, 1985

[36]Alan Gardner. *Egypt of the Pharaohs.* Oxford University Press, 1961, p. 79

[37]Said, op. cit., p. 128

[38]Said, op. cit., p. 137

[39]Edwards, op. cit., p. 98

[40]John Legon. 'A Ground Plan at Giza.' *Discussions in Egyptology,* vol. 10, 1988, pp. 33-39

[41]J. Davidovits. 'X-Ray Analysis and X-Ray Diffraction of Casing Stones from the Pyramids of Egypt, and the Limestone of the Associated Quarries.' *Science in Egyptology,* Manchester University Press, 1986, pp. 511-520

[42]Markson & Nelson, op. cit., p. 357

[43]M.A.A. Saayed et al. 'Quarternary Aquifer in the Nile Delta.' Annals of the geologicals survey of Egypt, p. 120

[44]M.M. Hassaan et al., op. cit., p. 229

[45]Zahi Hawass & Mark Lehner. 'The Sphinx: Who Built It, and Why?' *Archaeology,* vol. 47, No 5, 1994, p. 35

[46]Mark Lehner. 'The Development of the Giza Necropolis: The Khufu Project.' *Mitteilungen des Deutschen Archaeologischen Instituts Abteilung Kairo (MDIAK),* Vol. 41, p. 119, Springer Verlag, 1985

[47]PBS television special: *The Robot and the Pyramid*

CHAPTER 11: CONCLUSION

[1]Simon Welfare & John Fairley. *Arthur C. Clarke's Mysterious World.* A & W Publishers Inc., New York, 1982, p. 99

Appendix 1

Electromagnetic Energy Locations in the U.S., Canada, and England Open to the Public

This appendix lists some North American and English mound and rock chamber locations, as well as some sites of high geomagnetic energies. All are open to the public.

General Note: All the New England states except Rhode Island have fine examples of chambers, but these are on private land and not open to public access, with the fine and notable exception of Mystery Hill in North Salem, New Hampshire. New York has a few chambers that offer access to outsiders. You can access sites in New England through pre-scheduled trips with a number of organizations. For information on how to do this, we suggest logging on to the web site of the New England Antiquities Research Association (neara.org) who conduct a number of outings each year, with each outing visiting several chambers. They are also a useful link to other organizations and individuals who do the same.

The information provided is current as of the publication of the book. Sites sometimes change owners or contact organizations. Although the sites listed are usually in national parks or near roadways, use the same common sense when visiting the sites as you would when visiting any outdoor location. See individual site listings for particular hazards.

ALABAMA

Moundville Archaeological Park

Here was a 300-acre, Mississippian Culture village on Black Warrior River, protected on three sides by a wooden palisade. Inhabited c. 1000-1450 AD. There are twenty-six mounds, one with a restored "temple" on top.

Location: 14 miles south of Tuscaloosa, on Hwy 69 South
Hours: Daily: 8 AM to 8 PM
Museum open 9 AM to 5 PM
Fee: Adults $4, Native Americans free
Phone: 205-371-2234
Email: mbeo@bama.ua.edu (for special events)
Website: moundville.ua.edu/home.html

Oakville Indian Mounds Museum & Park.

The site contains the largest Woodland Culture mound in Alabama, 27-feet high, base covering 1.8 acres, top one acre. It was built by Copena Indians c. 2,000 years ago.

Location: 1219 County Rd. 187, Danville, AL 35619
Hours: Monday to Friday, 8 AM to 4:30 PM
Saturday to Sunday, 1 PM to 4:30 PM
Fee: Free
Phone: 256-905-2499
Email: HoytCagle@aol.com
Website: lawrencecounty.ala.nu/mound1.htm

ARKANSAS

Parkin Mound, Parkin Archeological State Park

In this park is a seventeen-acre village inhabited c. 1300-1550 AD. One large, flat-topped mound of the Mississippian Culture is preserved near St. Francis River.

Location: At junction of U.S. 64 and Arkansas 184 North, Parkin, AR
Phone: 870-755-2500.
Hours: Tuesday to Saturday, 8 AM to 5 PM
Sunday, 12 noon to 5 PM
Fee: Adults $2.75
Email: parkin@arkasas.com
Website: cast.uark.edu/~shelley/html/parkin/parkinmoundpg.html
arkansasstateparks.com/parks/park.asp?id=40

Toltec Mounds State Park

A large and complex mound site is located on the bank of Mound Pond. It once had an eight- to ten-foot earthen embankment on three sides. Formerly there existed sixteen mounds, up to fifty feet high, but several have disappeared.

Location: At #1 Toltec Mounds Road in Scott, AR. It is 16 miles southeast of North Little Rock and 9 miles northwest of England, Arkansas off U.S. Highway 165 on Arkansas Highway 386. Easy access from I40 interstate, take Exit 169 south on Arkansas Highway 15 for 14 miles to Keo, then northeast on U.S. Highway 165 for 4 miles; or take Exit 7 (England) going southeast on U.S. Highway 165 for 10 miles.
Phone: 501-961-9442
Hours: Tuesday to Saturday, 8 AM to 5 PM
Sunday, 12 noon to 5 PM
Fee: Interpretive center is free, tour of the site for "small admission fee"
Email: taltecmound@arkansas.com
Website: www.cast.uark.edu/~shelley/html/parkin/toltecvisitpg.html

CALIFORNIA:

The Ojai Foundation

A spiritual retreat and educational center located atop a small plateau that straddles the San Andreas Fault. If you're worried about earthquakes, you can rent a teepee for a night. We measured some puzzling and dramatic energy changes there. We have also seen photos by Ed Sherwood showing light balls at night there.

Location: 9739 Ojai-Santa Paula Rd., Ojai, CA 93023
Phone: 805-646-8343
Hours: Tuesday to Sunday, 9 AM to 5 PM
Fee: Free
Email: contact@ojaifoundation.org
Website: ojaifoundation.org

Mt. Shasta

This dormant volcano in northern California is still a center for purification rituals and sweat lodge ceremonies at Panther Meadow on the mountain. Numerous private, local groups offer vision quests on and around the peak, often at the many caves, springs, and smaller volcanic cones in the area (including Black Butte). Erupting every sixty years or so, Shasta still has a couple of active small fumaroles on the peak, one with an acidic hot spring inside it. While we have not taken measurements on this enormous "stratovolcano" at 14,160 ft. is likely to be as littered with magnetic anomalies as Bear Butte, due to its many past lava flows, etc.

Location: 60 miles north of Redding and south of the Oregon border next to Interstate 5
Phone: 530-926-3696
Hours: All year, daily
Fee: $15 adult summit pass, $5 recreational parking pass
Email: visit@ mtshastachamber.com
Website: mtshastachamber.com

Mt. Tamalpais

Paul Devereax reports the presence of a highly magnetic "seat" on a rock ledge. The magnetic anomaly is so strong it is probably the result of a lightning strike. Devereau says that wear patterns on the rocks suggest that many people have sat there over time, possibly in order to alter their consciousness.

Location: Just north of San Francisco in Mt. Tamalpais State Park
Phone: 415-258-2410
Hours: Winter, 7 AM to 6 PM
Summer, 7 AM to 9 PM
Fee: $6 per vehicle
Email: john@mttam.net
Website: mttam.net

CONNECTICUT

See General Note: Connecticut has rock chambers on private land. You can access them through tours run by the New England Antiquities Research Association: neara.org.

Gungywamp Swamp

Located near Groton, is a site of stone rows and of highly magnetic rocks, among them The Cliff of Tears. In this area we had very high electrostatic readings (page 28).

Location: The Gungywamp Society, P.O.Box 592, Colchester, CT
Phone: N/A
Hours: N/A
Fee: N/A
Email: gungywamp@gungywamp.com.
Website: gungywamp.com

FLORIDA

Lake Jackson Mounds Archaeological State Park

This site contains seven mounds, built c. 1000-1450 AD. The largest one measures 278 by 312 feet at base, height 36 feet.

Location: 3600 Indian Mounds Rd., Tallahassee, FL
Phone: 850-922-6007
Hours: Daily, 8 AM to sunset
Fee: $2 per vehicle up to eight people, $1 each additional person
Email: shirley.deal@dep.state.fl.us
Website: floridastateparks.org/lakejacksonmounds

GEORGIA

Etowah Indian Mounds Historic Site

This 54-acre, Mississippian Culture site contains six mounds, a 63-foot high, flat-topped earthen platform, and a defensive ditch. Its population counted several thousand c. 1000-1550 AD.

Location: 813 Indian Mounds Rd., S.W. Cartersville, GA
Phone: 770-387-3747
Hours: Tuesday to Saturday, 9 AM to 5 PM
Sunday, 2 PM to 5:30 PM
Fee: $2.50–$4
Email: etowah_mounds@mail.dnr.state.ga.us
Website: gastateparks.org/info/etowah

Kolomoki Mounds State Historic Park

This site has seven mounds built 350–950 AD by Woodland Culture people.

Location: 205 Indian Mounds Road, Blakely, GA
Phone: 229-724-2150
Hours: Museum opens daily 8 AM to 5 PM
Fee: Museum is $2.50 for adults

Email: Kolomoki_Park@mail.dnr.state.ga.us
Website: www.georgiaplanning.com/history/Kolomoki

Ocmulgee National Monument

This site contains mounds from two periods. Woodland Culture people (1000 BC-900 AD) built stone effigy mounds and earthen mounds. Early Mississippian Culture people (900-1150 AD) made huge earthen mounds and earth lodges.

Location: 1207 Emery Hwy, Macon, GA
Phone: 478-752-8257
Hours: Daily, 9 AM to 5 PM
Fee: Free
Email: N/A
Website: nps.gov/ocmu

ILLINOIS

Cahokia Mounds State Historic Site

This site bears remains of the most sophisticated prehistoric civilization in the U.S. (Plate 22). This Mississippian Culture city was inhabited c. 700-1400 AD and had a population of about twenty thousand. Originally there were more than 120 mounds of which 68 are preserved. Don't fail to visit the "Woodhenge," which are tree trunks stuck in the ground in a circle so as to provide sightlines to determine solstices and equinoxes. As discussed on page 89, the principal mounds lie along axes of magnetic transition. The entire complex lies at the intersection between two large, homogeneous zones of differing geomagnetic field strength just up and down the road from it. These zones give consistent readings for miles in either direction away from Cahokia, but transition abruptly at the site.

Location: 30 Ramey St., Collinsville, IL
Phone: 618-346-5160
Hours: Wednesday to Sunday, 8 AM to dusk

Fee: $2 donation for adults is suggested
Email: cahokia.mounds@sbcglobal.net
Website: www.cahokiamounds.com

Rockwell Mound

The mound covers nearly two acres and is fourteen feet high. It was built about 200 AD by a Middle Woodland Culture people.

Location: 500 North Orange, Havana (in a city park)
Phone: 309-547-3721
Hours: Daily, 8:30 AM to 5 PM
Fee: Free
Email: info-dmm@museum.state.il.us
Website: museum.state.il.us/ismsites/dickson

INDIANA

Angel Mounds State Historic Site

Around 1100-1450 AD, a town, enclosed by a stockade, was inhabited by two to three thousand people of the Middle Mississippian Culture. Several mounds are preserved.

Location: 8215 Pollack Avenue, Evansville, IN
Phone: 812-853-3956
Hours: Tuesday to Saturday, 9 AM to 5 PM
Sunday, 1 PM to 5 PM
Fee: $4 adult
Email: curator@ angelmounds.org
Website: angelmounds.org

Mounds State Park

Here are ten earthworks from the Adena Culture, including the Great Mound, built around 160 BC.

Location: 4306 Mounds Road. Located off I-69 east of Anderson, IN
Phone: 765-642-6627
Hours: Daily, 7 AM to 11 PM
Fee: $5 gate fee, $1 daily fee
Email: N/A
Website: in.gov/dnr/parklake/properties/park_mounds.html

IOWA

Effigy Mounds National Monument

Covers 2,526 acres and contains 195 mounds of which thirty one are effigies of animals. The largest is Great Bear Mound, which measures 137 feet long and 70 feet across the shoulders and forelegs. The mounds are about 1,000 years old. They are placed on limestone. Apostol reported a biolocation reaction over the baby bear mound (the oldest) and a few hundred feet southeast, strong airflow from small holes in the limestone, suggesting a cave.

Location: 151 Hwy 76, Harpers Ferry, IA
Phone: 563-873-3491
Hours: Daily, 8 AM to 4:30 PM (extended hours in summer)
Fee: $3 Adult, $5 Vehicle
Email: efmo_superintendent@nps.gov
Website: nps.gov/efmo

Toolesboro Indian Mounds National Historic Landmark

Seven Hopewell Culture mounds, built between 100 BC and 200 AD, are on a bluff overlooking the Iowa River.

Location: Highway 99 Toolesboro, IA
Phone: 319-523-8381
Hours: Daily, 12 noon to 4 PM third weekend in May to Labor Day
12 noon to 4 PM on weekends from Labor Day to October 31
Fee: Free
Email: lccb@lccb.org
Website: www.iowahistory.org/sites/toolsboro/toolesboro_history.html

LOUISIANA

Poverty Point Mounds

These are among the oldest mounds in the U.S. from several hundred years BC. They sit atop a large gravitational and magnetic anomaly. Apostol's sensitive (Chapter 3) showed involuntary muscle twitches while being driven blindfolded across this anomaly. Traditional interpretation says that this mound was built before full time agriculture had entered North America. However, discoveries of the past few years have pushed the date of non-maize farming in North America well past Poverty Point's date of 1,200 BC (similar to the Olmec's San Lorenzo). A spectacular bird-shaped effigy mound rises 70 feet high and measures 700 by 640 feet. It is estimated to have taken five million man-hours to build! You can view a free video on www.ArchaeologyChannel.org.

Location: From I-20, take the Delhi exit and travel north of LA 17, east on LA 134, and north on LA 577
Phone: 888-926-5492 or 318-926-5492
Hours: Daily, 9 AM to 5 PM
Fee: $2
Email: povertypoint@crt.state.la.us
Website: www.crt.state.la.us/crt/parks/poverty/pvertypt.htm

MAINE

See General Note: Maine has rock chambers on private land. You can access them them through tours run by the New England Antiquities Research Association: neara.org.

MASSACHUSETTS

See General Note: Massachusetts has rock chambers on private land. You can access them through tours run by the New England Antiquities Research Association: neara.org.

MICHIGAN

Norton Mounds Group

This is a complex of seventeen mounds, the largest 100 feet in diameter and 15 feet high, created during the Hopewell Middle Woodland Period (100 BC-500 AD). Many mounds destroyed by the expansion of Grand Rapids.

Location: In a city park in Grand Rapids, between Indian Mounds Dr. and Interstate 196
Phone: 517-373-1630 (City of Grand Rapids Historic Preservation Society)
Hours: N/A
Fee: N/A
Email: N/A
Website: www.michigan.gov/hal

MINNESOTA

Indian Mounds Park

People of the Hopewell Culture built the first mounds here before 500 AD. Only six out of thirty seven remain.

Location: Atop Dayton's Bluff along Mounds Blvd., off Hwy 94 east of St. Paul
Phone: 651-266-6400
Hours: Call for more info
Fee: Free
Email: karen.clark@ci.stpaul.mn.us
Website: nps.gov/miss/maps/model/mounds.html

MISSISSIPPI

The Grand Village of Natchez Indians

This was the capital of the Natchez Indians in the late 17th and early 18th centuries. The site is discussed in this book on page 92. There are three platform mounds, a plaza and habitation.

Location: In Natchez, turn off U.S. Highway 61, known as Seargeant S. Prentiss Drive within the city limits, at Jefferson Davis Boulevard just south of the Natchez Regional Medical Center. Proceed on Jefferson Davis Boulevard to the entrance gate of the Grand Village.
Phone: 601-446-6502
Hours: Monday to Saturday, 9 AM to 5 PM
Sunday, 1:30 PM to 5 PM
Fee: Free
Email: gvni@mdah.state.ms.us
Website: mdah.state.ms.us/hprop/gvni.html

Pharr Mounds

This site covers about eighty-five acres, containing eight mounds from the Middle Woodland period (100-200 AD). The largest mound is eighteen feet high.

Location: On the Natchez Trace Parkway (milepost 286.7), 23 miles northeast of Tupelo
Phone: 662-680-4025
Hours: Daily, dawn to dusk
Fee: Free
Email: N/A
Website: www.cr.nps.gov/nr/travel/mounds/Pha.htm

Nanih Waiya Mound

It is a large rectangular mound, measuring 218 feet long, 140 feet wide, and 25 feet high. The mound was probably constructed during the Mississippian Culture period (1000-1600 AD). The site is discussed in this book on page 93.

Location: 15 miles northeast of Philadelphia, on State Hwy 21
Phone: 62-773-7988
Hours: Daily, dawn to dusk
Fee: Free
Email: N/A
Website: www.cr.nps.gov/nr/travel/mounds/nan.htm

Owl Creek Site

This site holds five Mississippian Culture platform mounds built 1100-1200 AD.

Location: In Tombigbee National Forest, 2½ miles west of Natchez Trace Parkway on Davis Lake Rd., milepost 243.1 on Natchez Trace Parkway, about 18 miles southwest of Tupelo
Phone: 662-285-3264 (Tombigbee National Forest)
Hours: Daily, dawn to dusk
Fee: Free
Email: N/A
Website: www.cr.nps.gov/nr/travel/mounds/owl.htm

Winterville Mounds

Spread over forty acres, this impressive group of mounds includes one of the tallest in America at 53 feet. Our fluxgate magnetometer survey of this site fits the normal pattern seen elsewhere. Large individual mounds were placed on detectable anomalies, and the site is at the intersection of zones of differing geomagnetic strength.

Location: It is located on Mississippi Highway 1, 6 miles north of the intersection of Highways 82 and 1 in Greenville, Mississippi
Phone: 662-334-4684
Hours: Daily, dawn to dusk
Museum hours: Monday to Saturday, 9 AM to 5 PM
Sunday, 1:30 PM to 5 PM
Fee: Free
Email: wmounds@mdah.state.ms.us
Website: mdah.state.ms.us/hprop/winterville.html

MONTANA

Chief Mountain

A solitary and dramatic tower of rock on the U.S. Canadian border that has long been used for vision quests by the Blackfeet. Made of 1.5 billion year old sedimentary rock, Chief Mountain is a solitary knob of one tectonic plate at its border with another. It thus qualifies as a conductivity discontinuity and shares this property with Silbury Hill and Avebury in England, which are located on solitary knobs of one aquifer where it borders on another. The mountain is easily accessible as it lies on one of the entrances to Waterton-Glacier National Park. This erosion-isolated remnant is geologically known as a Klippen and as such it ranks with the Matterhorn.

Location: On the eastern margin of the upper plate of the Lewis Overthrust, in Glacier National Park/ Waterton National Park
Phone: Chief Mtn. Portal: 403-653-3535 (Canadian side). 406-732-5572 (U.S. side)
Hours: The road is open seasonally.
June 1 to August 30: 7 AM to 10 PM
September 1 to September 30: 9 AM. to 6:00 PM
October 1 to May 31: gate is closed
Fee: $10 with vehicle, $5 on foot (Glacier National Park), $5 Adult (Waterton National Park)
Email: glac_information@nps.gov (Glacier National Park)
Website: www.glacier.national-park.com

NEW HAMPSHIRE

See General Note: New Hampshire has rock chambers on private land. You can access them them through tours run by the New England Antiquities Research Association: neara.org.

America's Stonehenge

A 4,000-year old megalithic site covers about thirty acres on Mystery Hill. Many stone walls contain large, shaped standing stones. We surveyed this site with a fluxgate magnetometer and found that, as elsewhere; the rock chambers were placed with a magnetic anomaly right outside the door.

Location: Take I-93 to exit 3. Follow Rt. 111 east for 4.5 miles. Watch for the sign just past the North Salem Village Shops. Turn right at the intersection with the traffic light (Mobil Gas will be on your left). Follow this road for one mile. Entrance is on the right.
Phone: 603-893-8300
Hours: January 1 to June 20: 9 AM to 5 PM,
June 21 to August 31: 9 AM to 6 PM,
September 1 to October 31: 9 AM to 5 PM,
November 1 to December 31: 9 AM to 4:30 PM
(An excellent museum with books and a short film)
Fee: $9 Adult
Email: info@stonehengeusa.com
Website: stonehengeusa.com

NEW MEXICO

Petroglyph National Monument

Occupying eleven square miles on three distinct sites just outside Albuquerque, it is the largest assemblage of rock art in North America. Fifteen thousand prehistoric images are carved into the edge of a seventeen-mile basalt escarpment here at the biggest conductivity discontinuity in North America. These images were usually carved while the artists were in trances. The authors measured dramatic electric ground current activity, as well as electric charge in the air, during a single day's visit.

Location: Located at 6001 Unser Blvd. NW, Albuquerque, NM
Phone: 505-899-0205
Hours: Visitor Center opens daily, 8 AM to 5 PM
Fee: Entrance free, parking $1-$2 per vehicle
Email: N/A
Website: nps.gov/petr

NEW YORK

Balanced Rock

A 90-ton boulder balances on three flat limestone slabs (Plate 21). We measured a strong magnetic anomaly directly under the rock (page 73). The site itself lies at the intersection of two L-shaped zones of differing magnetic field strength.

Location: At Rte. 116, North Salem
Phone: 914-669-0245 (North Salem Historic Preservation Commission)
Hours: N/A
Fee: N/A
Email: N/A
Website: northsalemny.org (town of North Salem website)

Clarence Fahnestock State Park

This park is the site of an old magnetite mine and is littered with strong anomalies. The rock chambers at Kent Cliffs (below) are located within a 10-minute drive of the park.

Location: In Carmel, NY take Taconic State Parkway to Rte. 301, then west on 301
Phone: 845-225-7207
Hours: Daily, sunrise to sunset
Fee: N/A
Email: N/A
Website: stateparks.com/clarence_fahnestock_memorial.html

Kent Cliffs

Location of two rock chambers.
Location:
A. Take the exit for Clarence Fahnestock State Park off Taconic State Parkway but drive east on U.S. Rte. 301, away from the park. A few miles later, after passing a Buddhist monastery, one chamber is located on a curve on the immediate left shoulder of the road. This is the chamber with the glowing arch shown in Plate 18. Park across the road and be quiet and respectful

as this site is on private land, belonging to the owner of the house behind the chamber. Exercise extreme care because when you step out of the door of this chamber, you are two feet from the lane of a busy highway with fast-moving cars on a curve.

B. Continue east on Rte. 301 past the above chamber where the road makes a sharp curve to the right across from Farmer's Mill Road. If the road runs along a reservoir and you see a deli on the right, you have gone too far. On the right hand side of this curve, a few feet below road level and hidden (except for the door) in the foliage, is another of the chambers we used in our seed experiments, which has a magnetic anomaly right outside the front door. This is private property, owned by a person across the road, so show proper respect and restraint. There is safe parking for one car just past the chamber on the right shoulder.

Ninham Mountain State Park, Kent, NY

This is the location of the rock chamber with the glowing ball shown in Plate 19, and which has had powerful effects on seeds. When you enter the park, drive up the mountain until a barrier across the road blocks further progress. Park in the parking area and walk past the barrier up the dirt road to the right. About 50-100 yards up on the right lies the chamber, well hidden in foliage, about 20 feet off the road. You will see a cement frame doorway, which was a later addition by the occupants of the house, whose foundation can still be seen just uphill from the chamber. This chamber is located at the intersection of a large, homogeneous zone of lower geo-magnetic readings downhill and a zone of higher magnetic readings, with many magnetic anomalies immediately uphill from the chamber.

Location: Continue on Rte. 301 about 2-3 miles and make a 270-degree left turn onto Local Route 41 North (Gypsy Trail Rd.) After about a mile turn left onto a dirt road (Ninham Mtn. Rd.) at the sign for Ninham Mt. State Park

Phone: N/A

Hours: N/A

Fee: N/A

Email: info@townofkent.org

Website: townofkent.org (citizen-run site for the town of Kent)

NORTH CAROLINA

Town Creek Indian Mound

Well-restored mounds are surrounded by a stockade.

Location: 509 Town Creek Mound Rd., Mt. Gilead
Phone: 910-439-6802
Hours: Tuesday to Saturday, 10 AM to 4 PM
Sunday, 1 PM to 4 PM
Fee: Free
Email: towncreek@ncmail.net
Website: www.towncreek.nchistoricsites.org

OHIO

Enon Adena Mound (Knob Prairie Mound)

This Adena mound is 40 feet high, with a 574 feet circumference at the base.

Location: Indian Mound Circle, Indian Drive, Enon, off Dayton Bypass on Rt. 675
Phone: 937-864-7080 (Heritage Society Office)
Hours: Tuesday, 1 PM to 4 PM
Wednesday, 1 PM to 4 PM and 6 PM to 8 PM
Saturday, 10 AM to 2 PM (Heritage Society Office)
Fee: N/A
Email: Info@enonhistory.org
Website: enonhistory.org/6928

Great Circle Earthworks (formerly Moundbuilders State Memorial)

This sixty-six acre site preserves a once vast system of ditch and bank earthworks. It contains a circular earthwork 1,200 feet in diameter with grass-covered earthen walls ranging from eight to fourteen feet high. In the center are three lower, connected mounds. It also contains a large octagonal earthwork.

Location: 1 mile SW of SR 16 on SR 79 in Newark
Phone: 800-600-7178
Hours: Daily, dawn to dusk July 1 to September 5, and April 2 to June 30
Closed September 11 to April 1
Fee: Free
Email: N/A
Website: ohiohistory.org/places/newarkearthworks/greatCircle.cfm

Hopewell Culture National Historical Park

This site contains earthworks that date from the Hopewell Culture, c. 200 BC-500 AD. Every mound over six feet high that we measured was placed directly atop a magnetic anomaly. The one linear mound has numerous anomalies along its ridge. The group of mounds is surrounded by a henge-like ditch.

Location: 16062 State Rte. 104, Chillicothe. Located next to the sprawling state prison complex
Phone: 740-774-1125
Hours: Grounds open from sunrise to sunset.
Visitor center open daily, 8:30 AM to 5 PM
June to August, 8:30 AM to 6 PM
Fee: $2 Adult, $4 Vehicle
Email: N/A
Website: nps.gov/hocu

Miamisburg Mound

The largest conical mound in Ohio and possibly in eastern U.S. was constructed during the Adena Culture (800 BC-100 AD). It measures 877 feet in circumference and was originally more than 70 feet high. This mound was measured by us with a magnetometer. It sits directly atop a magnetic anomaly and is located at the intersection of two nested, L-shaped regions of differing magnetic strength.

Location: Mound Avenue, 1 mile south of exit 44 on State Rte. 725 and three miles west of exit 42 off I-75, in Montgomery County
Phone: 937-866-4532 (Miamisburg Parks & Recreation Dept.)

Hours: Daily, during daylight hours
Fee: Free
Email: N/A
Website: ohiohistory.org/places/miamisbg

Paint Creek

The entire area south of Chillicothe on U.S. Rte. 53, which runs along Paint Creek, is replete with numerous mounds, some of which are marked with signs and historical markers. This area was once home to large towns, and among the chief mounds are:

A. Fort Hill State Memorial
Location: Just down the road from Serpent Mound on Rte. 41 between Elmville and Cynthiana
Phone: 800-283-8905
Hours: Daily, May 28 to September 5, 2005 and
April 1 to June 30, 2006
Closed September 6 to March 31, 2006
Fee: Free
Email: N/A
Website: ohiohistory.org/places/fthill

B. Seip Mound State Memorial
This tall, linear mound displays magnetic variations across its top.
Location: On U.S. Rte. 50 between Bainbridge and Bourneville
Phone: 800-686-1535 (Ohio Historical Society Museum Division)
Hours: Daily, during daylight hours
Fee: Free
Email: N/A
Website: ohiohistory.org/places/seip

Serpent Mound

This quarter-mile long mound is the largest and finest serpent effigy in the U.S. Built atop a 100-foot-high bluff, it was constructed during the Fort Ancient Culture, around 1070 AD. The site is discussed in some detail on page 83.

Location: Serpent Mound is on State Route 73, 6 miles north of State Route 32 and 20 miles south of Bainbridge in Adams County
Phone: 800-752-2757
Hours: Tuesday to Sunday, 10 AM to 5 PM
Fee: $7 per vehicle. A modest museum with a good exhibit of the Unusual geology of the entire Serpent Mound region. Picnic facilities
Email: serpent@bright.net
Website: ohiohistory.org/places/serpent

Williamson Mound

This 28 ft. high, 156 ft. circumference Adena mound is set on a limestone ridge in Greene Co. in southern Ohio. Located in the Massies Creek limestone gorge, the only springs that can be observed in the one mile long gorge are three springs located about 500 ft. south of the mound. These were found by Apostol through a biolocation reaction from his sensitive (see page 24).

Location: From the Ohio Serpent Mound: Rte.73 North to Hillsboror, then Rte. 68 North to Xenia. From Xenia, Rte. 42 West to Cedarville. 1 mile south of Cedarville is the Indian Mound Park and the Williamson Mound
Phone: 937-562-7440 or 937-427-2883, ext. 7440 (Green County Park District)
Hours: N/A
Fee: N/A
Email: N/A
Website: co.greene.oh.us/parks/parks_&_facilities.htm (Green County Park District site)

OKLAHOMA

Spiro Mounds Archaeological Park

This site covers 140 acres and contains 12 mounds. The site was inhabited c. 850-1450 AD.

Location: 2½ miles east and 3½ miles north of Spiro on W.D. Mayo Lock & Dam Road off State Hwy. 9
Phone: 918-962-2062
Hours: Archaeological Center open Wednesday to Saturday, 9 AM to 5 PM Sunday, 12 noon to 5 PM
Fee: Free
Email: spiromds@ipa.net
Website: www.ok-history.mus.ok.us/mus-sites/masnum28.htm

SOUTH DAKOTA

Bear Butte State Park

We visited this volcanic mountain and found it littered with magnetic anomalies (page 33). It is also a very important site for religious ceremonies and vision quests of Native American tribes. Particular spots are marked with brightly colored 'flags' of cloth. (Plate 10)
Please observe that leaving the marked trails is prohibited.

Location: 6 miles northeast of Sturgis, off highway 79
Phone: 605-347-5240
Hours: All year, daily
Fee: $5.00
Email: BearButte@state.sd.us
Website: www.sdgfp.info/Parks/Regions/WestRiver/BearButte.htm

Harney Peak

The highest mountain in the Black Hills and long sacred to the Sioux and prior tribes, this was the only location where the authors themselves felt mind-altering influences. The source was electric ground current in a quartz vein in a cliff measured with our instruments. A well-maintained hiking trail of several miles leads to a lookout tower on top.

Location: Within Custer State Park, there are three trailheads to mark the beginning of trails leading to Harney Peak and into the Harney Range Trail System. Although these trails begin in Custer State Park, they enter the Black Hills National Forest and Black Elk Wilderness on their way to the summit.
Phone: 605-255-4515 (Custer State Park Headquarters), 605-673-9200 (Black Hills National Forest)
Hours: Daily
Fee: May 1 to Oct. 31, $5 per person or $12 per vehicle
Nov. 1 to April 30, $2.50 per person or $6 per vehicle (Custer State Park)
Email: CusterStatePark@state.sd.us (Custer State Park)
Website: www.sdgfp.info/Parks/Regions/Custer (Custer State Park)

Inyan Kara

The Sioux rank this mountain with Bear Butte and Devil's Tower. Like those it is an igneous intrusion, but unlike those it sprawls across 12 square miles. George Armstrong Custer climbed it in 1874 and carved his name on the top, which is still visible.

Location: In West part of Black Hills, Crook County, NE Wyoming, 14 mi S. of Sundance on private land near Newcastle Wyoming
Phone: N/A
Hours: N/A
Fee: N/A
Email: N/A
Website: wyoshpo.state.wy.us/inyan.htm

Sundance Mountain

Annual vision quest rites have been held here since time immemorial, hence its name. It is also the source of the alias of local criminal Harry Longabough, aka the Sundance Kid.

Location: In the Black Hills National Forest
Phone: 605-673-9200 (USDA Forest Service Black Hills National Forest)
Hours: Daily
Fee: Free
Email: r2blackhillswebinfo@fs.fed.us (Black Hills National Forest)
Website: www.fs.fed.us/r2/blackhills (Black Hills National Forest)

TENNESSEE

Pinson Mounds State Park

This 1,200 acre area contains 15 mounds and other earthworks from the Middle Woodland period (200 BC-500 AD).

Location: 460 Ozier Rd., Pinson. From Jackson, TN follow Hwy. 45 South to the small town of Pinson, TN. Turn left at the park sign, St. Rt. 197 and then follow the signs 2.5 miles to the park entrance
Phone: 731-988-5614
Hours: Museum open Monday to Saturday, 8 AM to 4:30 PM
Sunday, 1 PM to 5 PM
Fee: $3 per vehicle
Email: ask.tnstateparks@state.tn.us
Website: state.tn.us/environment/parks/parks/PinsonMounds

TEXAS

Caddoan Mounds State Historic Site

These mounds were built by the Caddoans, who lived here c. 800-1300 AD.

Location: On State Hwy 21, 6 miles southwest of Alto
Phone: 800-792-1112 or 936-858-3218
Hours: Thursday to Monday, 9 AM to 4 PM
Fee: $2 Adult
Email: N/A
Website: www.tpwd.state.tx.us/park/caddoan

VERMONT

See General Note: Vermont has rock chambers on private land. You can access them through tours run by the New England Antiquities Research Association: neara.org.

WEST VIRGINIA

Grave Creek Mound Archaeology Complex

This Adena mound, constructed about 250-150 BC, is the largest conical mound in the U.S. Its height is 69 feet and the circumference at the base 295 feet. It boasts a large museum focusing on the lives of the pre-Columbian Native Americans who inhabited the area and an extensive exhibit on the building of the mound. The mound is surrounded by a tall fence and can only be accessed during museum hours. It is located on a powerful magnetic anomaly, and we measured extremely strong earth currents there shortly after a thunderstorm had passed (page 85). Like so many mounds, it was originally connected to a river by ditch and mound avenues similar to the ones found at England's henges. The entire original complex was so striking that Lewis and Clark paused here for more than a day to explore it on their way to Missouri.

Location: 801 Jefferson Avenue, Moundsville, across the street from the state prison
Phone: 800-225-5982 or 304-843-4128
Hours: Monday to Saturday, 10 AM to 4:30 PM
Sunday, 1 PM to 5 PM
Fee: Museum requires a "small admission fee"
Email: gravecreek@wvculture.org
Website: www.wvculture.org/sites/gravecreek.html

WISCONSIN

Wisconsin is simply loaded with mounds, particularly in Madison, in and around the university campus. Most are effigy mounds, low structures made in the shape of an animal. They are also usually located at the edge of a lake. We have never measured an effigy mound and we do not know if these follow the magnetic siting pattern of the conical mounds we have measured in the Midwest. However, many of the parks cited below have modest conical mounds as well.

Avoca Mound Group

There are six linear and four conical mounds.

Location: In Lake Side Park, East Lake Shore Dr., Avoca, WI. The Avoca Mound Group is located in Lake Side Park and campground on the shore of Avoca Lake, a backwater slough of the Wisconsin River.
Phone: N/A
Hours: N/A
Fee: N/A
Email: N/A
Website: N/A

Azatlan Site

This is the premier archaeological site in Wisconsin. Between 1,000 to 1,200 AD it was home to Mississippian immigrants from Cahokia. Here, as at Cahokia, they built large earthen mounds and fortified the site with a huge timber and clay wall. Extensively excavated in the 1920's two platform mounds and parts of the wall have been reconstructed. To the northeast of Azatlan is a line of large conical mounds. A brochure for a self-guided walking tour is available on site.

Location: The park is on the east side of Jefferson County Highway Q, just south of County Highway B. Coming from the west on Interstate Highway 94, reach County B by going south on State Highway 89 to Lake Mills. Coming from the east on I-94, take State Highway 26 south to Johnson Creek to Highway B
Phone: 920-648-8774
Hours: Daily, 6 AM to 11 PM
Fee: $5 Vehicle
Email: Thomas.Davies@dnr.state.wi.us
Website: www.dnr.state.wi.us/org/land/parks/specific/aztalan

Baum Mound Group

The group holds three linear mounds and one conical.

Location: In Goodland County Park, Goodland Park Rd., Dunn, WI
Phone: 608-242-4576 (Dane County Parks)
Hours: N/A
Fee: N/A
Email: dane-parks@co.dane.wi.us
Website: countyofdane.com/lwrd/parks/parklist.asp (Dane County Parks site)

Burrows Park

This site features a bird effigy with 128 ft. wingspan.

Location: On Burrows Rd. off Sherman Ave., Madison, WI
Phone: N/A
Hours: N/A
Fee: N/A
Email: N/A
Website: N/A

Calmuet County Park

Six effigy mounds sit atop an escarpment over Lake Winnebago.

Location: County Trunk Highway EE, off State Hwy. 55 in Stockbridge, WI
Phone: 920-439-1008
Hours: Daily, sunrise to 11 PM
Fee: Free
Email: calumet.parks@co.calumet.wi.us
Website: www.co.calumet.wi.us/pk_home.htm

Cave of the Mounds

The mounds referred to here are natural limestone formations, in which a cave was discovered after blasting in 1939. It is listed here because Andrei Apostol's sensitive recorded a biolocation reaction while walking blindfolded over the cave (page 25). It is now a National Natural Landmark and open to the public in Dane County.

Location: Twenty miles west of Madison just off U.S. Highways 18 and 151 between Mount Horeb and Blue Mounds, WI.
Phone: 608-437-3038
Hours: **Spring:** *March 15 up to Memorial Day Weekend*
Weekdays, tours on the hour, 10 AM to 4 PM
Weekends, tours in thirty minute intervals, 9 AM to 5 PM

Summer: *Memorial Day Weekend to Labor Day Weekend*
Weekdays and Weekends, tours in twenty minute intervals, 9 AM to 6 PM

Fall: *Tuesday after Labor Day Weekend to November 15*
Weekdays, tours on the hour, 10 AM to 4 PM
Weekends, tours in thirty minute intervals, 9 AM to 5 PM

Winter: *November 16 to March 14*
Weekdays, tours on the hour, 10 AM to 4 PM
Weekdays, hours vary, call for tour times

Fee: $12 Adults, $6 Children (age 4-12)
Email: N/A
Website: caveofthemounds.com

Devils' Lake State Park

Low effigy mounds surround the lake, particularly on the north shore. The Bear Mound is located just above the pocket aquifer that maintains the lake at a constant level.

Location: In Baraboo, Wisconsin and situated along the Ice Age Trail
Phone: 608-356-8301
Hours: Daily, 6 AM to 11 PM
Fee: $10.00
Email: hesedr@dnr.state.wi.us
Website: devilslakewisconsin.com

Edgewood Mound Group

Twelve mounds are preserved on the campus of Edgewood College. A linear mound and six conical mounds are visible along Edgewood Drive. Two linear mounds are near the library, and on the other side of the library is a large bird effigy. Two conical mounds are near the playground.

Location: Edgewood College, Wooodrow St. (off Monroe Ave.), Madison
Phone: 800-444-4861
Hours: N/A
Fee: N/A
Email: N/A
Website: www.edgewood.edu (Edgewood College website)

Edna Taylor Conservancy

There are six linear mounds. Two are very long linear mounds that follow the crest of the hill. The four other mounds are on the northeastern slope of the drumlin.

Location: At Edna Taylor Conservancy, Femrite Dr., Madison
Phone: N/A
Hours: N/A
Fee: N/A
Email: N/A
Website: N/A

Elmside Park

This park has two effigy mounds: a bear and what has been referred to as a lynx.

Location: At Lakeland and Hudson Avenues, Madison
Phone: N/A
Hours: N/A
Fee: N/A
Email: N/A
Website: N/A

Farwell's Point Mound Group and Mendota State Hospital Group

Two large mound groups are on the shore of Lake Mendota. Farwell's group has a number of large conical mounds, part of a linear mound, and a bird effigy – built over a 1,000 year span! To the east lies the Mendota State Hospital group, with some of the finest effigy mounds anywhere, and several conical mounds. One bird mound has a wingspan of 624 feet! To see mounds check with the staff in the administration building, which is the first building on the right after you enter the hospital grounds.

Location: On Troy Drive in Madison
Phone: N/A
Hours: N/A
Fee: N/A
Email: N/A
Website: N/A

Governor Nelson State Park

Covering 422 acres on the northern shore of Lake Mendota, this park has five conical mounds. They were probably built during the Middle Woodland Period, while a large panther or water spirit effigy was a later addition. Also remnants of a stockaded Late Woodland Culture village are present.

Location: 5140 County Trunk Hwy M, Waunakee
Phone: 608-831-3005
Hours: Daily, 6 AM to 11 PM
Fee: $10 day pass per vehicle
Email: N/A
Website: wisconsinstories.org/2001season/native/nj_journey.html

Lake Koshkonong Effigy Mound Groups

Two hundred seventy mounds in twenty-three groups surround Wisconsin's third largest lake. Some lie atop the highest yielding parts of the aquifer that feeds the lake, but some lie atop the lowest yielding sections.

Location: In Jefferson Co., a few miles Southwest of Ft. Atkinson
Phone: N/A
Hours: N/A
Fee: N/A
Email: N/A
Website: N/A

Lizard Mound County Park

This park is named for its effigy mound shaped like a gigantic lizard and contains twenty-seven other effigy mounds, built 500-1000 AD.

Location: In West Bend, WI. On County Trunk 'A', one mile east of State Hwy 144. On Hwy 45 take exit STH 33
Phone: 262-335-4445
Hours: Daily, April 1 to November 15
Fee: Free
Email: webplan@co.washington.wi.us
Website: www.visitwashingtoncounty.com/parks.html

Outlet Mound

This is a large conical mound overlooking the outlet of Lake Monona. It is probably 2,000 years old.

Location: Midwood and Ridgewood Avenues, Monona
Phone: N/A
Hours: N/A
Fee: N/A
Email: N/A
Website: N/A

Picnic Point Mounds

On the south shore of Picnic Point, halfway to the tip of the peninsula, are two linear and three conical mounds. Near the tip is a conical mound.

Location: Willow Dr., University of Wisconsin, Madison
Phone: N/A
Hours: N/A
Fee: N/A
Email: N/A
Website: N/A

University of Wisconsin Arboretum

A bird and panther effigy mounds, plus linear and conical mounds are located on both sides of McCaffrey Road at the University Arboretum. One group is situated right above several prominent springs. A map is available at the McKay Center at the arboretum.

Location: The Arboretum's 1260 acres border the southern half of Lake Wingra. Vehicles can enter in 2 places: from the north at the intersection of McCaffrey Drive, North Wingra Drive and South Mills Street or from the south at the intersection of McCaffrey Drive and Seminole Highway, just north of the Beltline (Highway 12)
Phone: 608-263-7888
Hours: Trails open year-round

Fee: Free
Email: info@uwarboretum.org
Website: uwarboretum.org/visit

Upper Wakanda Park Mound Group

Three large oval mounds on a ridge overlooking Lake Menomin, a widening of the Great Cedar River.

Location: Wakanda Park, Pine St., Menonomie
Phone: 715-232-1664 (Menomonie Recreation Dept.)
Hours: Daily, 7 AM to 10 PM
Fee: Free
Email: tourism@menomoniechamber.org (Menomonie Chamber of Commerce)
Website: menomoniechamber.org/tourism/parks.asp (Menomonie Chamber of Commerce website)

Wyalusing State Park

This 2,628 acre park contains 21 mound sites, once totaling more than 130 mounds, from the Middle and Late Woodland Periods. 69 have been preserved, including the 'Procession of the Mounds' – a row of conical and linear mounds and one effigy that follows the crest of a bluff.

Location: 13081 State Park Lane, Bagley
Phone: 608-996-2261
Hours: Daily, 6 AM to 11 PM
Fee: N/A
Email: brian.hefty@dnr.state.wi.us
Website: wyalusing.org

WYOMING

Devil's Tower National Park

This geological twin of Bear Butte, called Bear Lodge by surrounding tribes, is again today the center for religious ceremonies of numerous Native American tribes, some of whom mark particular spots with brightly colored 'flags' of cloth, as at the vision quest site of Bear Butte. Physical evidence shows the tower was used for vision quests through the 1930's. After a decades-long ban on the ritual use of Devil's Tower the National Park Service is now allowing Native American groups to obtain permits for group activities on land located away from the tower itself. But the brightly colored cloths and medicine bundles found at Bear Butte are now also appearing at the base of the tower itself.

Location: Visitors traveling east on I-90, exit at Moorcroft, WY. Visitors traveling west on I-90, exit at Sundance, WY. Take 14 north to 24, follow 24 north to Devil's Tower
Phone: 307-467-5283
Hours: Daily
Fee: $5 Adult or $10 Vehicle
Email: deto_interpretation@nps.gov
Website: nps.gov/deto

CANADA:

Chief Mountain

A solitary and dramatic tower of rock on the U.S. Canadian border that has long been used for vision quests by the Blackfeet. Made of 1.5 billion year old sedimentary rock, Chief Mountain is a solitary knob of one tectonic plate at its border with another. It thus qualifies as a conductivity discontinuity and shares a commonality with Silbury Hill and Avebury in England, which are located on solitary knobs of one aquifer where it borders on another. The mountain is easily accessible as it lies on one of the entrances to Waterton-Glacier National Park. This erosion-isolated remnant is geologically known as a Klippen, and as such, it ranks with the Matterhorn.

Location: It sits on the eastern margin of the upper plate of the Lewis Overthrust, in Glacier National Park, MT, U.S. and Waterton National Park, Alberta, CAN
Phone: Chief Mtn. Portal: 403-653-3535 (Canadian side), 406-732-5572 (U.S. side)
Hours: The road is open seasonally:
June 1 to August 30, 7 AM to 10 PM
September 1 to September 30, 9 AM to 6 PM
October 1 to May 31, gate is closed
Fee: $10 per vehicle, $5 Adult (Glacier National Park)
$5 Adult (Waterton National Park)
Email: glac_information@nps.gov (Glacier National Park)
Website: glacier.national-park.com

Dreamer's Rock

This vision quest site on Manitoulin Island, Ontario is still used by First Nations people. Located in northern Lake Huron, Manitoulin is the largest fresh water island in the world and is a geological wonderland. Here the pre-Cambrian shield outcrops at the shorelines, forming a huge conductivity discontinuity at the edge of the electrically conductive water. Dreamer's Rock is a 300 ft outcrop of white rock perched above the lake with spectacular views. Heavily vandalized in 1974, the site was rehabilitated by First Nations young people. Please respect the privacy of anyone you see here using the site.

Location: Located off Highway 6 (west of Sudbury) on Birch Island where Highway 6 connects Manitoulin Island to the northern mainland. Tours are available in nearby Whitefish River and other towns. From southwestern Ontario, connection can be made up Route 6 north to Tobermory and Cape Herd, where a ferry connects to Route 6 on southern Manitoulin. Tribal permission must be obtained to visit Dreamer's Rock alone or with a tour. One place to obtain permission is at the Great Spirit Circle Trail organization, which is owned and operated by tribe members.
Phone: 1-877-710-3211 (toll free) or 705-377-4404 (Great Spirit Circle Trail Organization)
Hours: N/A
Fee: N/A
Email: circletrail@circletrail.com (Great Spirit Circle Trail organization)
Website: circletrail.com (Great Spirit Circle Trail website)

Majorville Medicine Wheel

The site is located in the Valley of the Ten Peaks in the Canadian Rockies near Calgary, Alberta. This is a major medicine wheel site that stayed in use for 5,000 years! It was first marked out about the time the builders of Silbury Hill in England cut their first chalk. Unfortunately it is extremely far off the nearest road and on private property, as are most of Alberta's medicine wheels. With an estimated two-thirds of all the medicine wheels in existence, Alberta is the center of this cultural phenomenon. Those wishing to visit sites, though, need to do a great deal of advance research, including obtaining permission from land owners. The Alberta Wilderness Association periodically hosts guided tours to the Majorville Medicine Wheel.

Location: In the Valley of the Ten Peaks in the Canadian Rockies near Calgary, Alberta, on the highest hill in the area
Phone: 1-866-313-0713 (toll free) or 403-283-2025 (The Alberta Wilderness Association)
Hours: N/A
Fee: N/A
Email: awa@shaw.ca (The Alberta Wilderness Association)

Website: albertawilderness.ca

Peterborough Petroglyphs

This site contains an impressive set of rock carvings that have been covered over by a well-designed building to preserve them and allow viewing in all weather. Petroglyphs Provincial Park is located ninety minutes NE of Toronto and 55 km NE of Peterborough. It contains 900 petroglyphs of turtles, snakes, birds, humans, and geometric shapes cut into white marble rock over 1,200 years ago. This is normally associated with vision quest activity. Much of the best art has been enclosed in a building with skylights. Handicapped accessible.

Location: Off Northey's Bay Rd., 11 km from Hwy. 28
Phone: 705-877-2552
Hours: Open seasonally
Fee: $2 Adult, $7 to $15 Vehicle
Email: N/A
Website: ontarioparks.com/english/petr.html

Petroglyph Park

On the east central coast of Vancouver Island, British Columbia, just across from the city of Vancouver. Two hectares hold hundreds of Native American rock carvings. The visible sandstone layers may provide a natural conductivity discontinuity. We have not surveyed this location.

Location: Take the main highway south out of Nanaimo. The park will be on the left (seaward) side just before the mouth of the Nanaimo River empties into the channel
Phone: N/A
Hours: During the operating season, park gates open at 7 AM and close at 11 PM
Fee: $3 to $5 per day Vehicle
Email: N/A
Website: britishcolumbia.com/ParksAndTrails/Parks/details/?ID=450

St. Victor Petroglyphs Provincial Historical Park

The park is located on a rock outcrop on an isolated hill with panoramic views of surrounding plains in Saskatchewan. It is a possible conductivity discontinuity and/or vision quest site though we have not yet measured it. Much of the rock art can be viewed from a wood walkway leading to another park on top of the hill.

Location: From Maple Leaf Highway 1: 40 miles west of Regina, take Route 2 south through Assinoboia. A little way past Route 719, watch for sign for left turn (east). Road bends 90 degrees south before park and another 90 degrees east just past park. Watch for signs.
Phone: N/A
Hours: N/A
Fee: Free
Email: N/A
Website: www.se.gov.sk.ca/saskparks/ParkInfo

Stein Valley

Located in British Columbia, Stein Valley is a traditional vision quest site where hikers can see faded rock paintings, presumably created by shamans to illustrate their trance visions. Preserved "forever wild" as a Native American cultural site. A 40 mile long roadless valley of true wilderness, for the experienced backpacker only (pack in all supplies but water and include a topo map, compass, and first aid kit). Recommended hiking time to traverse is one week, but the eastern end of the trail, which follows the Stein River, near Lytton is far easier than the strenuous west end near Stein Lake.

Location: Only 3 hours (100 mi.) from Vancouver. Take Hwy. 1 to the village of Lytton. Alternate route: From N. Vancouver take Hwy. 99 (the Sea to Sky Highway) to Pemberton Valley/Mt. Currie area. Trailhead at Lytton is at Van Winkle Flats sandy beachhead
Phone: N/A
Hours: N/A
Fee: N/A
Email: N/A
Website: wlapwww.gov.bc.ca/bcparks/explore/parkpgs/steinvly.htm

Writing-on-Stone Provincial Park

The park contains spectacular stone outcroppings in a genuine prairie ecosystem of 4,400 acres. It is home to the largest collection of rock art in the Great Plains with over fifty sites holding thousands of figures. We have not visited this site but from the rock outcroppings viewed, we suspect this may be a substantial conductivity discontinuity, and thereby qualitatively similar to Petroglyphs National Monument in New Mexico, where we have measured powerful electric ground currents. Bear in mind that most rock art was carved by shamans in trance.

Location: 19 mi. east, 4 mi. south of Milk River, Alberta. From Milk River east 32 km on Secondary Road 501, south 10 km on Secondary Road 500, paved
Phone: 877-877-3515
Hours: Daily, year round; Rock art tours, seasonal
Campground does not take reservations
Fee: N/A
Email: N/A
Website: www.uleth.ca/vft/milkriver/wospp.html

Vision Quest Stone Rings

This site is at Bannock Point, Whiteshell Provincial Park, Manitoba. We have not yet surveyed this location, but it's many stone "rings" in varying shapes and designs laid out on bedrock remind us of those stone rings that lay atop magnetic anomalies on Bear Butte and seem to serve as vision quest sites there. They are still occasionally used as sites for visions quests and rituals. Please respect the privacy of any Native Americans you see using the rings. Camping, cabins, beach, playground, even a meteorite lake.

Location: From Winnipeg, take Highway 1 for 126 km/78.3 mi. east to Falcon Lake and West Hawk Lake. Other entry points to the park include PR 307 at Seven Sisters Falls and PTH 44 at Rennie
Phone: 1-888-482-2267 (toll free)
Hours: Seasonal
Fee: $5 for three days per vehicle
Email: N/A
Website: gov.mb.ca/conservation/parks/popular_parks/whiteshell

ENGLAND:

Adam's Grave Long Barrow

It measures 200 feet long by 20 feet high. This burial mound has stones at one end and is surrounded by a ditch circa 3,000 BC. Our electrostatic voltmeter was "fried" here one afternoon by a mysterious mass of electric charge circling the hill as described on page 122. Open to the public.

Location: (SU113634) on Walker's Hill, 3-3/4 miles NE of Pewsey (A345, B3087), 1 mile NE of Alton Barnes.
Phone: N/A
Hours: N/A
Fee: N/A
Email: N/A
Website: N/A

Avebury Henge

This is one of the largest such structures in Europe. It has a 1400 ft. diameter bank and ditch with four causeways. The ditch is fifteen feet deep and forty feet wide. Ninety-eight standing stones remain in a circle and linear avenue. We took approximately one thousand magnetometer measurements here. This site is discussed in detail in Chapter 9. Museum and gift shop with books.

Location: (SU103700) 6 miles west of Marlborough. Just off the A4, the A4361 runs right through the stone circle. Open to the public during daylight hours.
Phone: +44 (0) 1672 539425 (Museum)
Hours: Museum open daily, Mid March to November, 10 AM to 5:30 PM
Saturdays, 1 PM to 4:30 PM
Sundays, 11 AM to 4:30 PM
Site is open year-round
Fee: Free
Email: ATIC@kennet.gov.uk (Avebury Tourist Info. Center)
Website: visitkennet.co.uk/avebury

Overton Hill (The Sanctuary)

The Sanctuary was originally the beginning of Avebury's West Kennet Stone Avenue. Watch for marker at road's edge. Modern stone markers depict the double stone ring that was erected here circa 3,000 BC. On a hill overlooking the River Avon, we measured a magnetic anomaly at the centre of the circles. If you bring a magnetometer to take measurements, beware of the buried water main that runs along this side of the road. It gives powerful readings and can easily confuse you.

Location: On the A4 half a mile east of Avebury, 6 miles west of Marlborough
Phone: +44 (0) 870 333 1181
Hours: Daily
Fee: Free
Email: customers@english-heritage.org.uk
Website: www.english-heritage.org.uk/server/show/conProperty.308

Silbury Hill

The hill is flat on top with spectacular views of surrounding countryside and the surrounding ditch and causeways. Several years ago a pit opened on the top when the ground collapsed that had been originally excavated during the twentieth century, though the 5,000-year-old sections are still holding up well. Traditionally a fence blocks off access to the mound with a sign "No Visitors Allowed," and everyone simply climbs over the fence and sign. This may or may not have changed since our visit. People frequently sit up through the night there. We conducted a multi-instrument one week monitoring of this site, which is discussed in detail in Chapter 9.

Location: Located half a mile from Avebury on the A4, 6 miles west of Marlborough
Phone: +44 (0) 870 333 1181
Hours: Daily
Fee: Free
Email: customers@english-heritage.org.uk
Website: www.english-heritage.org.uk/server/show/conProperty.310

Stonehenge

One of the wonders of the world. Erected between circa 3000 BC and 1600 BC. A circular structure, aligned with the rising of the sun at the midsummer solstice.

Location: 90 miles west of London on Salisbury Plain. Located at the junction of the A303 and A344/A360 2 miles west of Amesbury.
Phone: 0870-333-1181
Hours: Daily, 9 AM to 5 PM
Note: During the Summer Solstice, the site may be closed to the general public as thousands show up dressed in white robes for rituals.
Fee: £5.50 per adult, £2.80 per child
Email: customers@english-heritage.org.uk
Website: www.english-heritage.org.uk/stonehenge

West Kennet Long Barrow

A restored long barrow, reminiscent in structure of the stone chambers of America and Europe as well as the passage graves of France. We recorded a moderate geomagnetic transition here.

Location: Visible just off the A4 a few hundred meters east of Silbury Hill. ¾ mile SW of West Kennet along footpath off A4 (OS Map 173; ref SU 104677)
Phone: +44 (0) 870 333 1181
Hours: Daily, daylight hours only
Fee: Free
Email: customers@english-heritage.org.uk
Website: www.english-heritage.org.uk/server/show/conProperty.323

Windmill Hill

This ancient causewayed enclosure sits uphill from and behind Avebury, just off the Ridgeway public walking path. The site itself is nondescript and easily missed. Ask directions at the Avebury museum or the Red Lion Pub in the village located within Avebury stone circle. The site is discussed in Chapter 9.

Location: 1½ mile NW of Avebury (OS Map 173; ref SU 086714)
Phone: +44 (0) 870 333 1181

Hours: Daily
Fee: Free
Email: customers@english-heritage.org.uk
Website: www.english-heritage.org.uk/server/show/conProperty.324

Appendix 2

Do It Yourself!
(with your own instruments)

For those of you who wish to visit ancient sites and take measurements of your own, a wide variety of instrumentation exists. However, we have seen great change in this field since we began our investigations. Some of the models we have ourselves used have either been discontinued or superseded. Some are still available. Overall, they fall into three categories:

Ground electrodes. These also include a long length of wire and a voltmeter which is capable of registering millivolts of DC charge. The entire assemblage is quite reasonable in cost, probably under $200.

Fluxgate magnetometer. Our model measures the z-axis of the geomagnetic field only. This is the most pertinent of the x, y, and z axes. Such models are available for around $4,000 or less. At the time when we began our work, a model that measured all three axes cost about $30,000. Today, however, such models are available for a small fraction of that price.

One advantage of the more costly 3-axes units is their ease of operation. You basically just push a button when you want a reading. The z-axis units require some patience, dexterity, and practice. For an accurate reading, you must hold the probe parallel to the geomagnetic field line, while facing north. This is how you will obtain the highest reading. If you fail to hold the probe parallel to the field line, you will get lower readings. Thus, it is easy to mistakenly get a lower-than-accurate reading, but virtually impossible to get a higher-than-accurate reading. It requires standing still and rotating the probe in your hand while you watch the readout. With practice and a steady hand, you can get readings accurate to within about 50 gammas (or 0.5 milligauss) during a few minutes of manipulating the probe.

Increases of more than a few hundred gammas are likely to be non-natural sources, such as buried or nearby iron or steel objects. Don't worry about change or a pen in your pocket. I hold the probe at about chest height with bent elbow simply because it is the most comfortable. Choose a comfortable position because you will be holding it for a long time. If you need to work in the rain, you can put the unit's base, with readout, inside a clear plastic bag and just let the probe get wet.

Some z-axis units now on the market are accurate to as little as one gamma, but it is impossible to hold the probe steady enough to accurately measure to such fine resolution. If you wish to do so, you need to employ a clamp or base similar to a camera or telescope tripod. However, as we have seen, the coarser resolutions still can produce very useful information at most sites.

Electrostatic voltmeter. This is the unit that measures electric charge in the air or on objects. It can be used while walking around a site. You can also stand still and sometimes watch the readings change as gusts of wind deliver air that is more ionized. Today, a variety of types are available for under $1,000. One important aspect of use is to not wear clothing made of synthetic fibers, silk, or wool as the cloth rubbing against itself will produce electrostatic charge that can register on the meter. All cotton clothing is best, though leather is also fine. Rubber- or leather-soled shoes are best.

Normally, interference from overhead electric lines is not a problem. It will be a problem getting accurate readings in rain. You cannot put the unit in a plastic bag or under a nylon umbrella and get accurate readings because both generate electrostatic charge. Always store the unit in a cloth bag or leather case, never in a plastic or nylon bag, unless it is shielded by a cotton or leather covering. If you need to check if it is working, a good test target is your own hair after rubbing it with silk or rubber. That can produce a charge of several thousand volts.

E-mail us at the address below if you would like more detailed information, including:

- A how-to manual for using the instruments on site
- Links to manufacturers
- Pricing
- Our opinion of the detailed pros and cons of the various types of units

E-mail address: seedofknowledgestoneofplenty@yahoo.com.

Glossary

Aerosol – a cloud of fine droplets of minute size

Andesite – basaltic stone of a type found primarily in the Andes; moderately magnetic

Anomalous – literally 'unusual'; differing from it surroundings or from the norm

Anomaly – a spot with anomalous readings; i.e., a spot with higher or lower readings than the surrounding area

Aquifer – a layer of rock or sand that holds water

Basalt – volcanic stone, often magnetic

Brush discharge – occurs when an object with substantial electric charge electrifies the air around it strongly enough to give off a weak glow

Carbonaceous – of the Carbonaceous geological era; generally refers to limestone and chalk

Causeway – an undisturbed piece of ground that causes a break in a long ditch

Causewayed enclosure – concentric circles of such ditches with occasional causeways

Collider – modern tool of physicists, which accelerates sub-atomic particles to high speed and smashes them into atoms

Conductivity discontinuity – zone where a region of ground that conducts electricity well meets an area of ground that conducts electric current less well. Strong surges of geomagnetism and ground current often occur here

Corona discharge – used interchangeably with brush discharge

Crypto-explosion crater – a rare bowl-shaped geological formation, usually miles wide, caused by an enormous underground explosion of unexplained nature occurring eons ago

Diorite – a type of basaltic stone. The famous bluestones of Stonehenge are formed from this stone

Electrical conductivity – the ability of any substance to carry an electric charge across it. Water and copper have good electrical conductivity, while plastic and glass have poor electrical conductivity

Electrical ground current – natural DC electricity that runs through the ground

Electromagnetic energies – can manifest as electric or magnetic energy; one of the four basic forces of Nature, along with gravity, the strong atomic force, and the weak atomic force

Electrostatic voltmeter – a standard scientific instrument that measures the electric charge present on a surface. It can also be used to measure electric charge in the air; i.e., degree of ionization of the air

Emmer – primitive type of wheat, dominant in early European agricultural societies

Free radicals – usually oxygen molecules with a missing, or unpaired, electron. They are damaging to biological system, plant or animal, because they tend to try to rip the missing electron from cell wall membranes or anything else with which they come into contact. Accumulating free radical damage is the primary cause of aging. Anti-oxidants, such as Vitamin E work to combat free radicals in our own bodies

Gamma – a unit of magnetic measurement often also known as a nanotesla (nT). The earth's magnetic field is approximately half a gauss, fifty thousand gammas. By comparison, a simple refrigerator magnet can be dozens of gauss.

Geomagnetic field – the earth's natural magnetic field, believed to be produced by the rotation of molten iron deep in the planet's core as the Earth turns on its axis

Ground electrodes – specialized, spike-shaped sensors that are stuck in the ground at intervals and connected by wire through a voltmeter, which allows one to read the amount of electric current present in the ground

Henge – a ring-shaped ditch

Induction – principle of physics stating that an object moving through a zone of changing magnetic field strength acquires electric charge or current

Interfluve – a place on the ground's surface where two differing aquifer levels emerge and border one another. Interfluves are usually powerful conductivity discontinuities and, thus, spots with abnormally large daily surges of electric current and geomagnetic field strength

Island Effect – a term used in the study of electric ground currents (telluric currents) and their magnetic fields. An island will often exaggerate the normal daily fluctuations in both geomagnetic field strength and ground current, acting similarly to a conductivity discontinuity in this regard.

Ionization – electrification of a molecule; in this book, usually a molecule found in the air. This occurs when an air molecule such as oxygen obtains an unpaired electron and acquires negative charge, or when a molecule such as carbon dioxide loses an electron and, thus, acquires a net positive electric charge. The air is normally ionized in roughly balanced amounts of positively and negatively charged molecules. When the balance is tipped strongly in favor of either the negative or positive molecules, we say the air has become inoized

Ions – positively or negatively charged molecules, such as air molecules

Kelvin water dropper – method to measure electrical charge generated by water separating into two streams at a Y junction. The breaking of the surface tension that holds the water together in drops or streams causes some of the electric surface tension to be released

Khamsin – electrified wind from the Sahara. It blows seasonally for about fifty days in the spring. The word means 'fifty' in Arabic. It blows during the same period that the Nile was at its lowest levels, prior to the building of the Aswan Dam

Kokopelli – a mythological figure common in Anasazi rock art of the American Southwest

Magnetite – magnetic form of iron. It is not only attracted to a magnet, it acts as a magnet. It was called lodestone by the ancients because compass needles were made from it

Magnetometer – a standard scientific instrument measuring the strength of the magnetic field at any given place

Mastaba – rectangular Egyptian tomb built of quarried, cut stone blocks – the first structures in history to have been built from such stone

Megalith ('Large Stone') – structure made of large stones in pre-historic times

Menhir – French word for megalith

Meso-America – Mexico and Central America east to Panama

Neolithic ('New Stone') – a period of the Stone Age, c. 7000-4000 BC

Passage grave – a long, large form of rock chamber, common in France and elsewhere.

Peninsula Effect – a term used in the study of electric ground currents (telluric currents) and their magnetic fields. A peninsula will often exaggerate the normal daily fluctuations in both geomagnetic field strength and ground current, acting similarly to a conductivity discontinuity in this regard.

Plasma – the fourth state of matter (after liquid, solid, gas). In this book, usually a special type of electrically charged air mass. Air is often electrically charged, but when the behavior of a volume of air begins to be determined primarily by its electric charge, rather than such physical factors as wind, then it is classified as plasma. Lightning is an extremely high-energy plasma. There are many low-energy plasmas present in our atmosphere that do not glow

Pulsing – sudden, repeated changes of any type of reading

Pyramidion – stone of pyramid shape, placed atop Egyptian pyramids

Serpentine – basaltic stone, commonly of green color, moderately magnetic

Shaduf – old Egyptian method of lifting water up and into a raised field by means of a lever pole with a bucket on the end. This method had not yet been invented at the time the great pyramids were built

Telluric current – electric current running through the ground, near the surface

Tumulus – a round mound

Were-cat – fertility figure, half cat (the body), half human (the face)

Index

C

D

G

H

I

J

K

N

O

P

Q

R

S

T

V

W

Y

Z